*Lighthouses
& Keepers*

Lighthouses & Keepers

THE U.S. LIGHTHOUSE SERVICE AND ITS LEGACY

Dennis L. Noble

NAVAL INSTITUTE PRESS
Annapolis, Maryland

Naval Institute Press
291 Wood Road
Annapolis, MD 21402

First Naval Institute Press paperback edition, 2004
ISBN 1-59114-626-7

The Library of Congress has cataloged the hardcover edition as follows:
Noble, Dennis. L.
        Lighthouses and keepers: the U.S. Lighthouse Service and its legacy / Dennis L.
Noble.
        p. cm.
        Includes bibliographical references and index.
        ISBN 1-55750-638-8 (alk. paper)
        1. Lighthouses--United States--History. I. Title.
        VK1023.N63 1997
        387.1'55'0973--dc21
                                                97-20882

Printed in the United States of America on acid-free paper ∞
11 10 09 08 07 06 05 04   9 8 7 6 5 4 3 2 1

All photographs courtesy of the U.S. Coast Guard unless indicated otherwise.

Maps by Susan Browning

Frontispiece: The original Dry Tortugas light became a part of old Fort Jefferson, the
prison where Dr. Samuel Mudd, unjustly convicted of involvement in the assassination of
President Abraham Lincoln, was sent. In 1855, a new brick tower was built off
Loggerhead Key, and the light was moved to that location.

This book is for
Stacy N. Rose
A. J. "Joey" Rose III
Allie S. Noble
Joseph P. Noble
and
Kyle M. Ritten

Sometimes I think the time is not far distant when I shall climb these lighthouse stairs no more. It has always seemed to me that the light was part of myself. . . . Many nights I have watched the lights my part of the night, and then could not sleep the rest of the night, thinking nervously what might happen should the light go out. . . . I wonder if the care of the lighthouse will follow my soul after it has left this worn out body!

Abbie Burgess

CONTENTS

PREFACE

T HE MODERN U.S. COAST GUARD is the result of the merger of five
small federal maritime organizations. The oldest of the five, the U.S.
Lighthouse Service, is the one with which most Americans have an endur-
ing fascination: one writer even suggests that lighthouses are America's
answer to the castles of Europe. A Montana cowboy may not be able to tell the difference between a destroyer and a cruiser, but he more than likely can identify a
lighthouse. Why this allure? Perhaps because lighthouses represent a symbol of
steady permanence and a guiding light to a safe haven. Indeed, some organizations
use lighthouses in their logos for just these reasons. The light stations may also represent the pleasant times many families have spent by the seaside or their idyllic and
romantic dreams about the sea.

Whatever the reason for their captivation, lighthouse admirers owe a debt of gratitude to the U.S. Lighthouse Board. This group entered the picture in 1852, and
the condition of lighthouses, lightships, and buoys—at best, of poor quality—began
to improve. Eventually, this country became a leader in the field of aids to navigation through their efforts.

Throughout its long history, the system of such aids went by many names. To
prevent confusion, I will refer to the *organization* of aids to navigation as the U.S.
Lighthouse Service, which is better known than such other official titles as the U.S.
Light-House Establishment or Bureau of Lighthouses. I will also use the term *light
station* instead of lighthouse when discussing an individual unit, as that is the term
the service used. (Please see the glossary for a further explanation of this subject.)

While the number of books about lighthouses seems to grow each year, most
publications concentrate only on a single lighthouse or lighthouses within a geographic region; others are photographic essays. Most of the works do not show the
changes in aids to navigation brought about by technology, and, even more important, they do not treat lighthouses as a *part* of one of this nation's oldest federal

maritime organizations. Only two books, George R. Putnam's *Lighthouses and Lightships of the United States* (1913) and F. Ross Holland, Jr.'s *America's Lighthouses: Their Illustrated History* (1972 and subsequent editions), have attempted to put lighthouses in organizational perspective and to discuss technology. Holland points out that the majority of authors fail to use archival sources in their books; even Putnam's excellent book does not document sources. Historians wishing to delve into the National Archives' holdings, however, will be disappointed to discover that a large number of the documents were destroyed by fire, thus creating gaps in the official record.

A detailed history of the U.S. Lighthouse Service as suggested by Holland would be a massive tome. The work would have to include lighthouses, lightships, fog signals, buoys and buoy tenders, plus a study of the people who served in the U.S. Lighthouse Service. *Each* of the above aspects of aids to navigation should be the subject of a scholarly monograph. In addition, a history of the architecture of lighthouses is much needed. If all these studies were put into one volume, it would be unmanageable.

This book, therefore, is a one-volume overview, or *synthesis:* it briefly examines most of the aspects of the service, except architecture and river lights, from 1789 to 1939. It is meant to update Holland's and Putnam's works because of the recent appearance of a number of studies on the people who served at the lights, especially women. Holland and earlier writers also did not have access to such publications as *The Keeper's Log* of the U.S. Lighthouse Society, which provides a great deal of information on lighthouses and other aids to navigation. Furthermore, Holland did not cover buoy tenders.

Early in my writing, I became aware of the many characters who populate the story of the U.S. Lighthouse Service. In particular, four men—Stephen Pleasonton, Winslow Lewis, Augustin-Jean Fresnel, and George R. Putnam—greatly influenced the course of the service, and I have spent some time in chapters 1 and 2 outlining just how they did so.

Many readers may be disappointed that I did not cover more lighthouses. Instead, I have elected in chapter 3 to cover merely seven representative light structures to illustrate a given aspect of lights on the whole, such as a construction problem or location or type. Other lighthouses are mentioned throughout the narrative or in illustration captions, but not in depth.

A source of confusion for many readers who love the history of lighthouses is the height of light towers. There appears to be no accepted method of recording this fact: some authors measure from sea level (or the ground) to the base of the lantern room, others to the focal plane, and yet others to the top of the lantern room. Unless otherwise stated, I will use the height to the focal plane as recorded in the Light List.

Chapter 1 covers the beginnings of the service and ends in the pivotal year 1852, when the U.S. Lighthouse Board took control of the lights. Chapter 2 then covers the story of the U.S. Lighthouse Service from 1852 to 1939, when it was absorbed into the U.S. Coast Guard.

To me, the most interesting aspect of the lighthouses is the people who worked for the U.S. Lighthouse Service. Chapter 4 discusses the keepers, their duties, and their routines. My academic colleagues may look askance at chapter 5, which deals with ghosts and unusual tales—after all, how does one document a ghost?—but I feel that such stories give the reader an understanding of one side of a keeper's life at an isolated station. Some lighthouse employees complained that the trouble with their lives was that they had too much time to think. The stories within this chapter are a sampling of their thoughts.

An overview of lightships from their beginning to their demise is discussed in chapter 6. Life on a lightship was dangerous and lonely. I have treated the life of the crews on these lighthouses that went to sea in this chapter rather than put them with their brethren in shoreside lighthouses. The hard-working, but overlooked, buoy tenders and their crews are introduced in chapter 7. Fog signals, buoys, and electronic aids to navigation are introduced in chapter 8. I inform the reader of the modern ending of what was once the U.S. Lighthouse Service in chapter 9, which gives a brief background of lighthouses from 1939 to the summer of 1996.

When he was superintendent of lighthouses, George Putnam once wrote: "The lighthouse and lightship appeal to the interests and better instinct of man because they are symbolic of never-ceasing watchfulness, of steadfast endurance in every exposure, of widespread helpfulness. The building and the keeping of the lights is a picturesque and humanitarian work of the nation." I hope that readers will come away with an understanding of the service—but most especially the people, both good and bad—who inspired Putnam to pen his statement.

ACKNOWLEDGMENTS

I could not have written this synthesis of the U.S. Lighthouse Service without standing on the shoulders of two of the most important authors of the service: George R. Putnam, former superintendent of lighthouses, and F. Ross Holland, Jr., retired historian of the National Park Service. I also am indebted to the following people who helped me as I gathered material for the book.

While it is popular today to denigrate those who work in the federal government, the American taxpayer should be aware of how well that money is spent in the work of Dr. Robert M. Browning, Jr., the historian of the U.S. Coast Guard, and historian Scott T. Price. These two men are the *entire* heart of the U.S. Coast Guard's history program. Somehow they manage with professional skill to respond with courtesy and speed to a large number of requests and researchers. It continues to amaze me that an office so small can provide so many valuable services. I could not have completed this book without the help of either one.

Peggy Norris gave me her insightful comments, pointing out my inconsistencies and, of course, my misspellings. Tom Beard offered good suggestions. Truman R. Strobridge is extremely knowledgeable about the subject and pointed out areas where the manuscript needed improvement.

Wayne Wheeler, president of the U.S. Lighthouse Society, offered valuable insights plus the use of his organization's files. Capt. Gene Davis, U.S. Coast Guard (Ret.), and Larry Dubia, of the Coast Guard Museum Northwest, helped me with material on the lights of the Pacific Northwest. Researchers and those interested in the history of the U.S. Coast Guard should visit the Coast Guard Museum Northwest for an example of how volunteers can put together an outstanding museum. Ken Black, of Shore Village Museum, Rockland, Maine, gave me the benefit of his considerable expertise in aids to navigation. Richard J. Dodd, curator of Marine History at the Calvert Marine Museum, Solomons, Maryland, advised me on the Drum Point Light Station. Frank Ackerman, chief of interpretation, Cape

Cod National Seashore, provided much-needed information on the Cape Cod Light Station. Kevin Foster and J. Candice Clifford of the National Maritime Initiative provided material from their files.

Chuck Moser, of the Short Range Aids to Navigation Office, U.S. Coast Guard Headquarters, shared his vast knowledge of aids to navigation. The U.S. Coast Guard's Aids to Navigation Team, Port Angeles, Washington, likewise shared their insights on the automated lights they maintain. Cindee Herrick, curator of the U.S. Coast Guard Museum at the U.S. Coast Guard Academy, New London, Connecticut, provided records and assistance. Capt. B. W. Hadler and his staff at the U.S. Coast Guard's Seventh District Aids to Navigation and Waterways Management Branch gave me information about Sand Key.

Angie VanDereedt, archivist in Archives I, Reference Branch, National Archives, quickly provided the records on lighthouses and the U.S. Life-Saving Service that I needed. MacKinnon Simpson, of the Hawaii Maritime Center, assisted me with material on Makapuu.

I wish to thank freelancer Kim Cretors for her excellent editing and Scott E. Belliveau, J. Randall Baldini, and Linda W. O'Doughda of the Naval Institute Press for seeing the project through all of the stages of its production.

ABBREVIATIONS OF MILITARY RANK

Army

Gen.	General
Brig. Gen.	Brigadier General
Brvt.	Brevit
Col.	Colonel
Maj.	Major
Capt.	Captain
Lt.	Lieutenant
1st Lt.	First Lieutenant

Navy

Adm.	Admiral
Rear Adm.	Rear Admiral
Commo.	Commodore
Capt.	Captain
Comdr.	Commander
Lt. Comdr.	Lieutenant Commander
Lt.	Lieutenant
1st Lt.	First Lieutenant

*Lighthouses
& Keepers*

A Dim Beacon

I N THE ANCIENT WORLD, early mariners had little in the way of aids to
navigation to guide them safely into a strange port. One reason for this lack
of guidance is that many seaports feared the presence of a lighthouse would
enable enemy ships to locate and attack them more easily. The first lighthouse
structure in recorded history was the Pharos of Alexandria, one of the seven wonders of the ancient world. The Egyptians began their tower on Pharos, an island at
the entrance to Alexandria's harbor, around 300 B.C. and finished it about 280 B.C.
The structure saw service as a lighthouse for at least ten centuries.[1]

The Romans had several lighthouses at such locations as Messina, Ostria, and
Ravenna. Following Rome's fall, there is very little recorded about lighthouses during the so-called Dark Ages. By around 1100 A.D., trade began to flow more easily
between nations, and the shipping industry again was in need of aids to navigation.
The seafaring Italians led the way in lighthouse use. "Pisa built a light on the nearby
island of Meloria in 1157 and another near Leghorn in 1163; a tower was built in
1139, but not lighted until 1326; and a light was established at Venice about 1312."[2]

France and England soon had lighthouses. The French hold the honor of the
most elaborate structure since the Pharos of Alexandria. In 1584, Louis de Foix
began work on the light on the island of Cordouan at the mouth of the Gironde in
the Bay of Biscay. It took twenty-seven years to complete the project and then, to
the dismay of the builder, the entire island washed away. To save his work, de Foix
built a barrier around the light.

The description of this light almost staggers the imagination. The lower section
of the light, which contained the keeper's living quarters, measured 134 feet in diameter, and a central hall, 52 feet in diameter. A second floor, unbelievably, contained
a chapel. The next floor, at a height of 197 feet, held a giant lantern and chimney
designed for wood fires. So that keepers would not dirty the main part of the building when carrying wood to the fire, the inside spiral staircase was set off to one side.

Sandy Hook Light Station, New Jersey, is one of the oldest light towers in the United States.

The outside of the structure certainly was not utilitarian. Pillars, ornate windows, statues, and frescoes completed the light structure. The lighthouse of de Foix only partly survives today; the upper sections of the building were removed in 1788 and replaced by a circular stone tower sixty feet tall. "Until early this century, the Cordouan lighthouse was considered the finest in the world, and some say it still is."[3]

WHEN HENRY WINSTANLEY announced that he intended to build a lighthouse on Eddystone Rocks, some fourteen miles out to sea from Plymouth, England, "he was considered quite mad."[4] He anchored the structure to the rock with iron rods twelve-feet long, enclosing the upper part of the rods in a circular stone base twelve-feet high. Over the years, many lighthouse projects ran into construction delays, but the Eddystone has a unique excuse for work stoppage: in 1697 a French privateer captured the work crew and hauled them off to France where they remained prisoners of war. When finally released, the workers returned to their project.[5] The polygonal main building was of wood. It took four years to complete the project, and Winstanley, so confident of his design, said he wished to be in it "in the greatest storm that ever blew under the face of heaven." Winstanley received his wish. On 26 November 1703, a storm toppled the tower, killing Winstanley and some workmen.[6]

The next light tower on the rocks was under the supervision of John Lovett and John Rudyerd. Lovett supplied the money and Rudyerd the construction know-how. Rudyerd secured the foundation of the tower with iron bolts and then laid a base of stone and wood, with the main tower also of stone and wood and sheathed in wooden planks.[7] The tower went up in flames in 1755. The next builder of a lighthouse at the deadly rocks was John Smeaton. Smeaton's tower was in the shape of a cone, with a large base to give it a broader foundation. Unlike the other lights, Smeaton's creation was entirely of stone. The first light shone from this tower in 1759, and the tower stood for more than a century. In 1882, Trinity House decided

The old and the new at Cape Henry, Virginia. The original tower, at left, was completed in 1792 and the newer structure in 1881. The old tower stands 72 feet high, the new 165 feet above sea level.

Not all light towers were round, as illustrated at the Beavertail Light Station, Rhode Island.

the tower's foundation was no longer secure, and a newer, higher tower of 149 feet came into being. The "newer" light is still in operation. Instead of destroying Smeaton's work, workers removed every block of stone and reassembled the tower at nearby Plymouth.[8]

Even though the colonies depended on the maritime link to the Old World, it is interesting to note only an estimated seventy lighthouses existed in the Western world. Improvements in illumination developed slowly. Some of the first illuminated aids to navigation in the New England colonies were lighted baskets hanging from a pole atop a prominent hill.[9] The lighthouses of the individual colonies were erected near important ports of trade, mainly where people lived, and usually not close to major hazards to navigation. This is understandable given the times and conditions under which lighthouses were established and maintained. Local people, generally merchants, would appeal to the colony for a lighthouse. The colonial government would then construct the lighthouse in the area requested by the local residents. The reasoning went that local people built and maintained the tower, so why should they support a tower in the wilds of the country, such as Cape Hatteras, known as the "Graveyard of the Atlantic," or the Florida Keys?

Before the colonies broke with England, there were at least eleven permanent lighthouse structures in what is now the United States. Three other lights had been started and all the materials to begin construction on another, at Cape Henry, Virginia, had been purchased but work had not yet begun.[10] When the first permanent structure designed as a lighthouse was built in the United States has never been accurately determined. Most lighthouse studies, however, give the lighthouse on Little Brewster Island, in Boston Harbor, the honor of being the first lighthouse in North America. Boston's light is a good example of how light stations were established prior to 1789.

In 1713, local merchants petitioned the General Court of Massachusetts for a "Light Hous and Lanthorn on some Head Land at the Entrance of the Harbor of Boston for the Direction of Ships and Vessels in the Night Time bound into said Harbor."[11] The General Court appointed a committee to look into the need for a lighthouse; the committee agreed with the petition and recommended a light be established on small Beacon Island (now Little Brewster) at the entrance to the harbor. In June 1715, the court approved the petition and appropriated the funds for construction. To pay for the lighthouse and its maintenance, the court established light dues consisting of "one Penney per Ton Inwards and another Penney Outwards, except Coasters, who are to pay Two Shillings each, at their clearance Out, and all Fishing Vessels, Wood Sloops, etc. Five Shillings each by the Year."[12] Boston's light was first displayed on 14 September 1716.[13]

The fledgling U.S. government quickly realized the national value of lighthouses. The emphasis the federal government placed on aids to navigation is shown by the ninth law passed by the new government. On 7 August 1789, the central government assumed the responsibility for all aids to navigation and took over all existing

Boston Light Station is considered the first light station in the United States, but the present tower dates back to 1789, making it the second-oldest tower. It is the only manned light station remaining in the United States.

lighthouses as well as those under construction. (This act also marks the first provision for public works.[14])

Congress placed the financing of all aids to navigation in the Treasury Department. Unlike the financing of the lights in the colonies, there would be no light tax: the aids would be supported by appropriations from the general revenue. Alexander Hamilton, the first secretary of the treasury, personally oversaw the lights. When reviewing the correspondence of the early lighthouse service, one is struck by the involvement of highly ranked government officials in the operation of the service. Presidents George Washington, John Adams, and Thomas Jefferson personally approved lighthouse contracts and appointments to lighthouse positions. Of course, the government of the time was small, and there were few aids to navigation. As the number of aids grew and government became more complex, presidents could no longer find the time to administer as closely and the control passed more completely to the secretary of the treasury. In 1792, the office of the commissioner of revenue was created within the Treasury Department, and Hamilton moved the control of the aids to that office. The office was abolished from 1802 to 1813, and Secretary of the Treasury Albert Gallatin assumed direct control of the aids. In 1813, the secretary returned the aids again to the commissioner of revenue, who again took control until 1820, when the aids to navigation came under the control of the fifth auditor of the treasury. During this period, from 1789 to 1820, the number of lighthouses in the United States increased from twelve to fifty-five. One study of the lighthouses during this period noted that the lights seemed to have been built "to meet an immediate and pressing local need, without any reference to a general system."[15] Aids to navigation would remain under the control of the fifth auditor for the next thirty-two years, and this control would provide the most controversy in the history of the lighthouse service.

The fifth auditor was Stephen Pleasonton, who has been described as "zealous," "hard working," and an "overly conscientious guardian of the public dollar"—not altogether bad traits for an official who was one of the nation's principal bookkeepers. On the other hand, a former historian of the U.S. Coast Guard pointed out that Pleasonton became a "villain" in the history of the lighthouse service. He also brought to the job "the bookkeeper's lack of imagination."[16]

Pleasonton's responsibilities beyond the lighthouse service were great. He was responsible for "the diplomatic, consular, and bankers accounts abroad, and all the accounts at home appertaining to the Department of State and Patent Office, as well as those of the census, boundary commissioners, and awards of commissioners for

A stormy day at the Boone Island Light Station, Maine. It probably best represents how most Americans imagine a lighthouse.

An illustration of the light station at Cape Henlopen, Delaware, in 1767, which was threatened constantly by shifting sands. In 1926, the tower toppled during a storm.

adjusting claims on foreign Governments."[17] For a time, he also held the duties of the commissioner of revenue. Twelve years after Pleasonton took over the control of the lighthouse service, he had a total of nine clerks working for him. At that time, the lighthouse service consisted of 256 lighthouses, 30 lightships, and a number of buoys and beacons. Pleasonton had no maritime background or, for that matter, did he have any experience relating to maritime affairs. Perhaps this combined with the number of additional responsibilities that devolved upon him were mitigating factors for his poor stewardship of the lighthouse service.[18]

In the public's mind, Pleasonton was the general superintendent of the lighthouses. To administer the lighthouses, he appointed as his direct representatives the collectors of customs who had lighthouses in their districts. These collectors became superintendents of lights. The superintendents handled virtually all personnel matters, except for the actual appointment of the keepers, which was done by the secretary of the treasury. The collectors also selected and purchased the sites for lighthouses and oversaw the construction of the structures. They authorized the funds for maintenance, and they inspected the lighthouses annually. For their labors, the collectors received a 2½ percent commission on all lighthouse disbursements as an extra compensation.[19]

Pleasonton kept tight reign of the local superintendents, letting them have very little authority over the lights within their control. Prior to 1822, they had to clear all expenditures with the fifth auditor, and after that date he allowed them to spend up to $100 without approval. Despite this tight control, Pleasonton for some reason—probably his lack of expertise in maritime affairs—relied heavily upon the superintendents for information on the local lighthouses. If a complaint was filed against a light, Pleasonton would have the superintendent investigate; Pleasonton accepted "as virtually gospel whatever the local superintendent reported."[20]

Throughout his administration, Pleasonton pointed with great pride to the economy with which he operated the lighthouse service. In 1842, for example, he declared to Congress that he ran the lighthouse service of the United States at half the cost of the English lighthouse service. In 1851, the secretary of the treasury appointed an investigation into the lighthouse service, and the board gave Pleasonton high marks for zealousness in economy.[21]

During the Pleasonton years, most of the supply and work on the lights was done under contract. Oil and money to purchase other supplies were furnished to a contractor who made yearly visits to the lights. During construction of lighthouses, a supervisor was hired by the fifth auditor's office to oversee the work. It appears that a knowledge of construction was not necessarily a requirement for hire in this important duty. The first broad contract issued was to purchase Winslow Lewis' patent for a "reflecting and magnifying lantern."[22] (The relationship between Lewis and Pleasonton became a critical factor in the history of the lights. More on this later, but suffice it to say that Pleasonton tended to listen carefully to Lewis because he had maritime experience.) Lewis, as will be seen, had a vested interest in lighthouses. The record shows that Pleasonton's almost stubborn acceptance of Lewis's lighting apparatus was one of the reasons for the beginning of a poor reputation of the lighthouse service. One historian has noted that Pleasonton's reign of the lighthouses might best be labeled the era of "the lowest bidder."[23]

Congress began to be dissatisfied with the lights of the United States. In 1837, the first ripple in a growing wave of discontent over lights began, when the Treasury Department requested the legislature authorize a large number of new lighthouses. Before approving such a request, Congress decided to have a board of navy commissioners examine the sites and determine whether all the lights were actually needed. The commissioners found that thirty-one lighthouses could be eliminated.[24]

An act of 7 June 1838 divided the Atlantic Coast into six districts and the Great Lakes into two. The act specified that a naval officer be assigned to each district and that he should report on the condition of each aid to navigation within his district. Further, the naval officer should be the official to select sites for new lights.[25]

The newly appointed officers found the quality of the lights ran from extremely poor to quite good. Many lights were bunched together. On the Maine coast, for example, one officer reported that between two locations there were nine lights. At Nauset, there were three lights built 150 feet apart. The inspector noted that one light would have been sufficient. There were also structural defects in many lighthouses. The largest complaint, however, was in the quality of the light. At some loca-

Stephen Pleasonton, who had no maritime experience, controlled the lighthouses of this country from 1820 to 1852. He has been branded a "villain" for his poor stewardship during his tenure. (Library of Congress, courtesy U.S. Lighthouse Society)

tions the keepers went about their duties in a superficial manner, and some devoted very little time to the all-important duty of cleaning the lighting apparatus. Not every lighthouse had defects, but one historian of lighthouses notes that at least 40 percent had serious problems.[26]

For some reason, the national legislature did not act upon the 1838 report of an obviously flawed lighthouse service. In 1842, Congress appointed a select committee to look into the operation of several departments of the government with an eye to realignment, which would cut expenses and employment. In the Treasury Department, one of the areas under investigation was the office of the fifth auditor. At the same time, the Commerce Committee was investigating whether lighthouses could be operated more efficiently under the supervision of the Bureau of Topographical Engineers. In a move that seems to make no sense, and probably indicates that no one on the committees had any real maritime experience, it was found that lighthouses should remain within the Treasury Department and that Pleasonton was doing "tolerably well. . . ."[27] Helping the committee to arrive at this decision, Pleasonton appears to have exaggerated. For example, he claimed that there was a system for classification of lighthouses. In reality, there was no formal system. During the next major investigation in 1851, Pleasonton admitted to just that. Strangely, the fifth auditor was allowed to continue his stewardship of the lighthouses.[28]

On 19 June 1845, the secretary of the treasury had two lieutenants, Thornton A. Jenkins and Richard Bache, detailed from the navy and sent abroad to study the lighthouses of Europe with the goal of learning how to improve the lights of the United States. The two officers submitted their findings, and the secretary reported the information to Congress. Congress failed, however, to act on any recommendations.[29]

As time passed, the growing wave of complaints about U.S. lighthouses reached tidal-wave proportions. Congress, as is their wont, moved slowly. In 1847, construction of six lighthouses was taken away from the fifth auditor and entrusted to the Corps of Engineers. Finally, four years later in 1851, the wave broke and engulfed the fifth auditor. Through the years, the chief detractors of the lights were Edward M. Blunt and George W. Blunt, who published the *American Coast Pilot*. This publication helped guide mariners in navigating the coast of the United States, and one important aspect of the guide was the identification of coastal lighthouses. For years, the Blunts received the complaints of mariners about the lighthouses, which they steadily passed on to Congress.[30]

Congress required the secretary of the treasury to appoint a board made up of two high ranking naval officers, two army engineering officers, a civilian "of high scientific attainment," and a junior officer of the navy to act as secretary to investigate all aspects of the aids to navigation in the United States. Secretary of the Treasury Thomas Corwin appointed as the investigating board Commo. William B. Shubrick, Comdr. S. F. DuPont, Bvt. Brig. Gen. James Kearney, Prof. A. D. Bache, and Lt. Thorton A. Jenkins.[31]

Secretary Corwin gave the board carte blanche authority to poke into every aspect of the nation's aids to navigation. The board carried out their assignment with a will—it is difficult to find an area that the group did not explore. Their investigative report filled an amazing publication running to 760 pages. Succinctly, the board found nothing right with the lighthouses of the United States or their administration. The board did acknowledge that the number of lighthouses had grown rapidly from 1820 to 1852 and that "great credit is due to the zest and faithfulness to the present general superintendent, and to the spirit of economy which he has shown."[32]

The board felt the only solution to the terrible condition of the lights was a complete revamping of the system. They suggested that a board consisting of groups of people representing several professions govern the lighthouses. This type of arrangement had proved successful in many of the European countries that had excellent reputations for aids to navigation. The board would be made up of nine members, with six in the pattern of the investigative board. The other three members would be an additional civilian scientist, another secretary from the army engineers, and the secretary of the treasury, who would be ex-officio president. Further, an inspector would be appointed to each district, and that official would be either an army engineer or a naval officer. The board would have complete control over the lights and would issue rules and regulations for the management of the service and instructions for keepers to properly run a lighthouse. A system for the classification of lights would also be devised by the new board. The board also urged the adoption of the Fresnel lens for use in lights.[33]

Congress, with little delay, signed off on the board's suggestions and on 9 October 1852 created the nine-member U.S. Lighthouse Board. Commo. William B. Shubrick and Comdr. S. F. DuPont were the naval members; Brig. Gen. Joseph G. Totten, of the Corps of Topographical Engineers, and Lt. Col. James Kearney, of the army engineers; A. D. Bache, superintendent of the Coast Survey, and Joseph Henry, first secretary of the Smithsonian Institution and a leading physicist, were the civilian members; Lt. Thorton A. Jenkins was the naval secretary; and Capt. E. L. F. Hardcastle was the engineering secretary. Commodore Shubrick was the chairman of the board, and the secretary of the treasury was ex-officio president of the board.[34]

Thus ended the thirty-two year lighthouse rule of Stephen Pleasonton. Using twenty-twenty hindsight, it is not difficult to see the fifth auditor was doomed to failure, and one can almost feel some sympathy for the man. In 1822, there were seventy lighthouses in the United States. Sixteen years later, the number had ballooned to 204 lighthouses and 28 lightboats, and in just another four years, there were 256 lighthouses, 30 lightboats, 35 beacons, and nearly 1,000 buoys. By 1852, there were 331 lighthouses and 42 lightships. Throughout this amazing period of growth, the fifth auditor did not have a professional organization made up of those familiar with maritime matters or engineers under his command. What he had working for him in Washington were clerical administrators who contracted for anything dealing with aids to navigation.[35] Furthermore, during Pleasonton's tenure, Congress was penny-pinching, and many lighthouse keepers were political appointees who had no great desire to tend the lights. Again, it is obvious that such an arrangement was doomed to failure.

No history of the lighthouse service fails to mention that Pleasonton was a conscientious, hardworking bureaucrat, whose first order of business was saving money for the government. Rear Adm. Charles Wilkes noted that Pleasonton "was as honest and upright a gentleman as ever served the country, exact in all his business transactions and with much judgement and bonhomie as could be found in very many." Wilkes, however, admitted the fifth auditor "was not a bright man, but with his accomplished and handsome wife they had much influence in the intricacies of the [government.]"[36] Pleasonton felt that one of the great accomplishments of his administration was that he could consistently claim to Congress that he returned more money back each year to the general fund than any other department. Herein lay Pleasonton's downfall. Like many who only see the "bottom line," the fifth auditor lost sight of his true job. He was responsible for providing the United States with good aids to navigation for those who sailed in American waters. The lives of sailors depended upon the quality of the aids. Pleasonton's thrift saved money, but he sacrificed the quality of the country's aids to navigation. The noted lighthouse historian F. Ross Holland, Jr., writes that "one wonders how many ships that wrecked during Pleasonton's thirty-two-year administration would have been saved had more effective lights been available."[37]

THE HISTORY OF LIGHTHOUSES is the story of a continual search for the best type of light for use in aids to navigation. Before beginning to trace this story, it may help to give a background of some of the terms and equipment that will be discussed in this and following chapters.

One of the ways a navigator may distinguish a lighthouse is by the light's characteristic. The light may be *fixed* (steady) or *occulating* (flashing). The light may be red or white. The characteristics of the earliest lights in the United States, because of the state of technology, generally were fixed, white lights. The French first perfected a means of rotating either the light and the reflectors or the lenses. The method was a simple clockwork motor, much like the works of a grandfather clock, with a descending weight activating the rotational movement. The lighthouse at Cape Cod, as early as 1798, had a device that enabled a screen to revolve around the oil lamps so that the light would be obscured at intervals, which produced a flash. When Fresnel lenses came into operation in this country, the large, heavy lenses rested on wheels or ball bearings in a track positioned at the base of these lenses. The clockwork turned the lens, which had its prisms and bull's-eyes arranged in panels. The light beam moved slowly, but the turning lens caused the viewer of the light to see a flash, with periods of dark between. The times of darkness and flashing provided the characteristic of the light. Several times a night keepers would have to wind the clockwork mechanism to raise the weight to the top of the tower in order to keep the lens turning. This seemingly simple arrangement could actually hold some unpleasant surprises. In 1869 a keeper at Maine's Petit Manan light reported that the large weights from the clockworks at the top of the tower came crashing down and fell to the lower floors with "such tremendous force that they snapped away eighteen steps of the cast-iron circular staircase." Luckily, no one was injured in this mishap.[38]

To create a red light, or any other color for that matter, colored panels were inserted into the lens, or the lamp had a solid color. The fixed light at Drum Point, Maryland, however, had three ruby-red panels attached to the inside of the lantern-room windows. This produced three sectors of red and one of white. (Because of Drum Point's location near land, the light needed to cover only 270 degrees.) A vessel wishing to navigate into the Patuxent River remained within the white sector of the light.[39]

The beam of light underwent many changes over the years. In 1727, at the great French Cordouan tower there was an attempt to intensify the light beam. A cone of wood covered with tin suspended point-down reflected the fire. In the 1730s, the

Swedes experimented with reflectors. William Hutchinson, an Englishman, in 1763, observed a parlor trick whereby a wooden bowl, lined with looking glass, reflected enough light from a single candle by which to read a newspaper two hundred feet from the light source. Hutchinson thought about the demonstration and then used small lamps placed in the middle of a reflector. One writer says this produced the "first really efficient lighthouse. . . ."[40] Eventually, the reflectors had a parabolic contour, a bowl shape that gathers rays of light and produces a concentrated beam. This became known as the *catoptric* (or mirror) system. When the Fresnel system came into use, this became the *dioptric* (or see-through) system.[41]

Both systems, however, needed the 1780 development of the Argand burner, "which became the first modern light source." Aime Argand, a Swiss, found that "an oil lamp with a cylindrical wick, confined within two concentric tubes and a glass chimney, produced a more brilliant light" than anything up to that time. "The French found that a three-wick Argand lamp provided as much light as 200 candles."[42]

There was no classification of lights in this country until after the adoption of the Fresnel lens. Thereafter, the order of the lens determined the classification of the light. Usually, the large first- and second-order lenses were for important coastal lights, and the other orders were for bays, harbors, and rivers. Thus, the important coastal light at Cape Hatteras became a first-order light because of its lens.

Geography played an important part in lighthouse design. If a light stood high atop a cliff, the height of the light structure could be relatively small, while a flat coastal plain would need a tower of some height. At Point Reyes, California, for example, the light is situated near the top of a steep cliff 225 feet above sea level, and the structure is only 23 feet from the base of the tower to the focal plane. At the other extreme, across the continent at Cape Hatteras, North Carolina, the light is sited in a low, sandy topography and therefore must be high: the focal plane of the light is 191 feet above high water.

Those interested in the light stations of the United States prior to the American Revolution will become quickly frustrated by the lack of information available on the structures. There are almost no construction details. Apparently, no two lights were built from the same set of plans, and all used local building materials. There are, however, a few details that are common to most of these structures. The lights prior to 1789 were built of wood or stone. Understandably, the towers built of wood fell victim to fire. The stone towers were generally built by piling stones one on top of another and held together by mortar. The walls do not appear to have contained any other reinforcement, but they were tapered as they rose so that the base could support the weight of the structure and also to prevent the tower from becoming unstable.[43]

The workmanship of the early lights varied greatly. Brant Point, on the south side of Nantucket Harbor, Massachusetts, was the second light in the United States and was first lit in 1746. The structure was cheaply built, and twelve years later was destroyed by fire. It was rebuilt in 1759 and toppled in a violent storm in 1774.

Again, the town replaced the light, which was once more destroyed by fire. A smaller light structure was built in 1783, but the light was so dim that a new beacon was built three years later. Another storm destroyed the light yet again in 1786.[44] On the other extreme of durability is the lighthouse at Sandy Hook, New Jersey. Sandy Hook is the only lighthouse now standing that was built prior to the American Revolution.[45]

New York merchants requested the establishment of the Sandy Hook lighthouse, which would help ships seeking New York City's harbor. To pay for the construction, two lotteries were established and, as in Boston, the cost of maintaining the structure was financed by light dues. The light was built by Isaac Conro and first displayed in 1764. The masonry tower was described as being "of an Octagonal Figure, having eight equal Sides; the Diameter at the Base 29 Feet; and at the top of the Wall, 15 Feet. The Lanthorn is 7 feet high; the Circumference 33 feet. The whole Construction of the Lanthorn is Iron; the Top covered with Copper. There are 48 Oil Blazes. The Building from the Surface is Nine Stories; the whole from Bottom to Top 103 Feet."[46]

Conro's structure was well built. During the Revolution, the Americans were afraid that the light would be useful to the British, so they tried to damage or destroy the tower. The tower was too well built and was not ruined. In 1852, the Sandy Hook lighthouse was described as one of the three best masonry light towers in the United States.[47]

After this basic background on the development of some of the equipment of lighthouse lighting, a discussion on illumination is in order. As mentioned, until the twentieth century, the basic ingredient of the light was a flame. The very earliest aids were open fires fueled by wood on hills to serve as beacons. When towers were eventually built, the fire on the hill was transported to an open area on the top of the structure. Even though this method was in use for centuries, there were many difficulties. Wood burned quickly and needed constant stoking. In the sixteenth century, there was a shift from wood to coal. This provided a bright light that could be seen for longer distances. Coal burned more slowly, thus keepers did not have to constantly monitor the fires. The problem with this source of illumination was that coal also tended to melt the fire grates. If the fire were enclosed with glass, the soot coated the glass. To increase the intensity of the light, engineers introduced reflectors, but these also became coated with soot.

At nearly the same time as coal was introduced into lighthouses, some lighthouses in Europe began to use candles. This source of illumination was not as messy as coal and was much easier to use. Furthermore, candles could be enclosed with lanterns to provide a steady light. The negative side was that they did not give off as bright a light. Despite this drawback, candles continued to be used in some lighthouses into the nineteenth century.[48]

The next step in the evolution of the light was the already mentioned introduction of lamps. The lamps tended to give off some smoke that coated the glass of the interior of the lantern and dimmed the light. To counter this, one new method was the creation of a lamp consisting of a pan of oil from which four wicks extended. Fittingly dubbed "spider lamps," they were first used in the United States at Boston light around 1790. The problem with the spider lamps was that they gave off acrid fumes, which burned the eyes and nostrils of the keeper and limited the time he could spend in the lantern room. Even with this drawback, spider lamps remained in use in the United States until at least 1812, when they were replaced by the next innovation in lighting.

The next important step in lighting aids to navigation came in 1781 with Amie Argand's lamp. The Argand lamp and parabolic reflector long had been in service in European lighthouses. The lamp and reflector became the foundation for the basis of the next major advancement in U.S. lighthouses and brought forth a man called "canny" and "crafty," with a personality described as "sanguine." Winslow Lewis was a sharp Yankee businessman, who was also "inventive and resourceful."[49] Winslow Lewis is the second major personality in the history of aids to navigation in this country, and he held center stage in the history of lighthouses in the United States for more than forty years.

Born 30 May 1770, the son of Winslow Lewis (a merchant captain), in the village of Wellfleet on Cape Cod, Massachusetts, the younger Lewis also followed the sea. At a relatively early age, on 3 October 1797, he was voted in as a member of the Boston Marine Society, which was founded in 1742 to promote the general interests of navigation.[50]

By 1805 the future looked bright for the thirty-five-year-old captain. He had just taken command of the 323-ton *Sally* of the Boston Importing Company, which operated ships regularly between Boston and Liverpool, England. Ashore, he had married and had six children. But then international events entered the picture. President Jefferson's embargo of 1807 put an abrupt halt to Lewis's promising career.

Forced to "swallow the hook," Lewis had to find something that would support his family. The former captain lacked the "rudiments of a formal education," but he was a sharp observer of his former maritime world. Like many a captain he had observed the poor quality of the lighthouses in this country. Why not improve the lights? A government contract could prove lucrative.[51]

Lewis began to experiment in 1807. The first test was in the cupola of the Boston state house. On 24 June 1808, Lewis obtained a patent for lighting the binnacles of ships and two years later, on 8 June 1810, he patented his "reflecting and magnifying lantern." Prior to receiving the patent, a committee from the Boston Marine Society received a request from Lewis to compare the quality of his light with others in the Boston harbor area. The committee was sufficiently impressed with the outcome that they wrote a glowing report on the results.[52]

Winslow Lewis's invention was a combination of devices that were already in operation. Lewis took the Argand lamp and combined it with the parabolic reflector and a lens similar to one developed by the English glass cutter Thomas Rogers in 1788–1789. This lens sat with the plane surface toward the lamp and a few inches in front of it. Lewis's lens was of molded, impure, slightly green glass, nine inches in diameter and two and one-half inches at its axis, and set in a copper rim. Even though the lens actually diminished the lamp, it was a large improvement over the spider lamps. The lamp and the reflector were the real reasons behind the success.[53]

In July 1810, Lewis received permission from the secretary of the treasury to test his device in one of the twin lights of Cape Ann, Massachusetts. The collector of customs, Henry Dearborn, who observed the test, was impressed with the results, but equally influencing him was the fact that Lewis's lamps used only half the oil as the lamps in the other tower. Dearborn wrote to the secretary of the treasury recommending that all lighthouses be equipped with this invention.

Secretary of the Treasury Albert Gallatin concurred with Dearborn and requested Congress authorize him to purchase Lewis's patent right away. He also wanted the Treasury Department to contract with Lewis to install the devices in all the then forty-nine lighthouses and keep them repaired for "a term of years not less than seven," at Lewis's expense. The amount of money appropriated was $60,000. Lewis also managed to convince the government that it would be cost effective to purchase a schooner, the *Federal Jack,* and fit it with a blacksmith shop, a carpenter shop, and bunking spaces for thirteen men along with the necessary equipment to carry out the work.[54]

From all indications, Lewis threw himself into the work. By the end of 1812, forty of the forty-nine lighthouses were converted. The War of 1812 intervened, preventing Lewis from completing his assignment. Finally, in the autumn of 1815, all U.S. lighthouses were equipped with Lewis's invention. Collector of Customs Dearborn wrote to the secretary of the treasury that now the lighthouses of the United States were "equal, if not superior, to any in the world."[55]

Lewis eventually branched out into other manufacturing and "became a man of affairs" in Boston. He also entered politics, while simultaneously keeping his interest in lighthouses. The commissioner of revenue, Samuel H. Smith, requested that the secretary of the treasury disregard the normal procedure of public advertising and bidding to maintain and supply the oil for the lighthouses.[56]

In 1816, Lewis entered into a contract with the government to supply the oil for the lighthouses and to annually visit each lighthouse in person to maintain the lamps and to report on the condition of the lighthouses to the Treasury Department. As one student of this period of lighthouse history has noted: "the contract made Winslow Lewis the *de facto* superintendent of lighthouses."[57]

Lewis made the most of this arrangement and soon began determining such things as the number of lamps at a light and the diameter of the reflectors used at a station. In 1817, Lewis published the first U.S. light list. In some cases, complete sailing directions were given, much like the Blunt's *American Coast Pilot.* Because his lamps used less oil, there was a surplus in the amount of oil contracted for, and

Lewis was free to sell off or use this remainder. Lewis's biographer, Richard W. Updike, has brought out that in 1816 the profit from this surplus could have run as high as $12,472.38. This was "more than twice the amount of the Secretary of the Treasury's salary" and almost half the U.S. president's salary. When Pleasonton took control of the lights, he reduced the amount of oil Lewis received, but the amount the contractor received for transportation and maintenance of the lights was increased—Lewis continued to have a lucrative contract. Not too surprisingly, complaints about Lewis's conduct began to circulate. In 1819, David Melville accused Lewis of stealing and patenting an idea of Melville's to keep lighthouse oil warm in the winter.[58] It was implied that Lewis was more interested in keeping oil as an illuminant than experimenting with gas in order to continue his large profit margin. In 1827, the five-year contract ended and a New Bedford firm underbid Lewis. Lewis shifted his attentions to the building of lighthouses.[59]

Lewis had already entered into lighthouse construction in 1818, undertaking a contract for a lighthouse on Frank's Island on the Northeast Pass of the Mississippi River. Knowing the structure would be difficult, he accepted the contract on two conditions. First, the government would employ an inspector to insure that Lewis followed the plan correctly and, second, if the foundation gave way before the completion of the project, he would still be paid. The government agreed and awarded an $80,000 contract that used the plans by the noted architect Benjamin Latrobe. Lewis's request for the two conditions were based upon his opinion that the plan was "injudicious." He publicly stated that the foundation would not hold the weight of the planned lighthouse. Three days before the project was completed, the foundation failed. In 1822, Lewis submitted his own plan for the lighthouse and, using some of the salvageable material from the former structure, built a tower for $9,750.[60]

It was Winslow Lewis's invention and his experience with lighthouses that apparently caused Pleasonton to rely so heavily upon Lewis's opinion. Given Lewis's record, and Pleasonton's lack of expertise, this is not too surprising. By 1821, Pleasonton was using Lewis as a guide to accepting the cost of erecting a lighthouse. Lewis often would say a contract for a light was too high, the bid would be rejected, and Lewis would then be awarded the contract. This caused Lewis to become the principal builder of lighthouses in the United States. Lewis won so many contracts that he drew up a set of plans for the five different sizes of towers that he believed would meet the needs of any land location.[61]

Not content with cornering the market on constructing lighthouses, Winslow Lewis became the leading contractor for refitting old lights. Lewis apparently felt that he had, in fact, become the head of the lighthouses. In 1835, without any direction from Washington, he changed the characteristics of the Mobile Point lighthouse. To change the characteristics of a light without consulting and warning mariners is a major step and should be accomplished slowly. Lewis received a "rebuke" from Pleasonton, but when a number of maritime interests indicated the change was unwise, the fifth auditor came to Lewis's defense.[62]

All this seems to point toward corruption; many nineteenth-century critics concluded just this, but the charges were never proven. One of Lewis's major critics, strangely enough, was a nephew, Isaiah William Penn Lewis. The young Lewis was a sailing master in the Caribbean trade and seemed enamored by his uncle's success in the lighthouse field. After the age of thirty, Isaiah left the sea and studied civil engineering. Lewis employed Isaiah in the planning and inspection of lighthouses. Eventually the ambitious Isaiah decided to strike out on his own and became a competitor against his uncle in the lighthouse field. He obtained several contracts, but on some he bid too low and tried to have the fifth auditor compensate him for his losses. Pleasonton would definitely turn a deaf ear to this plea. According to one student of Winslow Lewis, once this happened, Isaiah Lewis "became an implacable enemy of the lighthouse establishment and his uncle. . . ."[63]

In 1842, Isaiah stated that his uncle's invention had actually been copied from an English lighthouse and that his reflectors were bad. The nephew also accused the uncle of incompetence in constructing lighthouses. The attack was very personal. Isaiah even "asked Secretary [Walter] Forward how Fifth Auditor . . . Pleasonton came to head up America's lighthouse system." Pleasonton came to Lewis's defense stating that Lewis's work was satisfactory. Lewis published a rebuttal and, basically, called Isaiah incompetent to judge his work. Like that of his nephew, Lewis's attack became very personal. The most consistent critics of Lewis, however, were the Blunts.[64]

Winslow Lewis died 20 May 1850, just prior to the major change in the lighthouse service. One obituary stated that "his name would be long held in respect and veneration by all who have business on the great ocean."[65] It would, however, be more accurate to say that Lewis's name has become more than tarnished in close to a century and a half since those words were penned. Detractors point out that his lamp and lens were not really that good. Arnold Burges Johnson, chief clerk for the lighthouse board, wrote in 1890 that Lewis's reflectors "came about as near to a true paraboloid as did a barber's basin."[66] Furthermore, added his critics, many of his lighthouses crumbled.

Lewis's biographer, Updike, however, notes that Rear Adm. George C. Remey, chief of the lighthouse board in 1904, pointed out the Lewis reflectors were so well made that in certain places "it is now a question whether a better light was possible than was then furnished." Furthermore, George R. Putnam, the long-serving commissioner of lighthouses (see chapter 2), wrote that "it is true that many of these lighthouses were later rebuilt more substantially, yet the class of work done probably met at a moderate cost the immediate needs of a growing country."[67]

From the available evidence, it appears that Winslow Lewis and Stephen Pleasonton did not commit criminal acts. Pleasonton, however, can be accused of favoritism, and this favoritism stems from the fifth auditor's lack of knowledge of maritime affairs. Perhaps understandably, Pleasonton turned to a man, Lewis, who appeared to be very knowledgeable about all phases of lighthouses and also had a strong maritime background. Lewis, as a sharp Yankee businessman, did everything he could to insure that his business would continue. What the two men are guilty of is stopping the improvement of the lighthouses of the United States. Pleasonton slavishly adhered to the low-bidder principal, while, again, Lewis did what he could to keep his business going.

What this did was to effectively prevent the introduction of the French Fresnel lens into this country, which would be the major innovation in lighthouses until well into the twentieth century. As early as 1830, Pleasonton knew of the lens, but, despite pressures from Congress to adopt the lens, he held off. In the meanwhile, Pleasonton, because he respected Lewis's knowledge, allowed Lewis to obtain a monopoly over something the government wanted. Lewis would of course prefer to keep this type of arrangement and would do nothing to help introduce something new into lighthouse design. There seems little question that Pleasonton held off from adopting the Fresnel lens because of Lewis. In the end, Pleasonton and Lewis can be accused of helping retard the development of lighthouses in the United States. One should not lose track of the fact that such a situation worked against

The light from the Fire Island Light Station, New York, pierces the night.

This bold black-and-white diamond pattern marks Cape Lookout Light Station, North Carolina.

those that go down to the sea in ships and added an additional danger to a profession already fraught with many hazards. It is not too much to say that until both Stephen Pleasonton and Winslow Lewis were out of the picture, the United States could not have a first-class lighthouse service.

One of the largest improvements in aids to navigation came about because of the work of the third pivotal person in the history of American lighthouses. This man, who did not have the personality of Pleasonton or Lewis, overcame great obstacles to greatly influence lighthouses, yet remains little known to many. It is not hyperbole to state that the obscure French physicist Augustin-Jean Fresnel (pronounced "Fraynel") revolutionized the light emitting from lighthouses.

Fresnel was born in 1788 in Broglie (at the time called Chambrais) to Jacques Fresnel and Augustine Merimee. Augustin-Jean was the second of four sons. The family moved to the Norman village of Mathieu in 1789, where Fresnel spent his boyhood. Poor health plagued him throughout his life. His teachers thought the frail and quiet boy slow-witted when, at the age of eight, he could not read and had difficulties remembering his daily lessons. All his life Fresnel hated "even the simplest exercises of memory." When Fresnel entered secondary schooling, however, he began to exhibit a remarkable ability for mathematics. At sixteen, this ability won for him the annual competition for entry to the Ecole Polytechnique. While his scores in some subjects were dismal, those in graphic arts and geometry "surprised and delighted the examiner."[68]

The students of the Ecole Polytechnique, under the influence of the Napoleonic era, wore uniforms. Napoleon wanted the best students trained to be officers. Those not strong enough for military service were relegated to be civil servants and engineers used for building roads and canals. With his poor health, Fresnel was assigned to finishing his technical training at the school of bridges and highways.

After graduation, he soon learned to despise constructing bridges and highways. "There is nothing I loathe more than having to lead men," he once said. Fresnel, however, performed his duty, received promotion, and was sent to the Rhone Valley on a major road project.[69]

During his free moments of supervising road crews, Fresnel began to think about light. At this time, the accepted scientific view of light was that formulated 150 years previously by Isaac Newton—namely, that light consisted of a swarm of "corpuscles" moving through the "ether." Newton also thought each light "corpuscle" had a different mass that varied according to its color. He also decided that light "corpuscles" in the form of rays were reflected or refracted as they pleased. Some were reflected and others preferred refraction. He explained this as "fits." Newton's ideas had been accepted for more than a century and became an almost impossible barrier for Fresnel to break.[70]

The parabolic reflector and the bull's-eye lens then in use in lighthouses was known as catoptric. The problem with this system was light tended to escape from both the top and sides of the device. This led to the refractive lens, which bent light and directed it to a desired point. This is known as dioptric. What Fresnel accomplished was a single lens known as a *catadioptric* that combined the two former systems of focusing light. The lens contains a central reflecting bull's-eye surrounded by a series of concentric prismatic rings and refracting prisms. In very brief terms, light is directed by the prismatic rings to the central bull's-eye where it emerges as a single concentrated shaft of light traveling in one direction. Fresnel's lens, in general, is unchanged today.[71]

The amazing thing about Fresnel's work with light was how little he was accepted in his time. Although receiving notice from some of the leading scientific personalities of the time—such as André M. Ampère, the French physicist who helped establish the science of electrodynamics—still he was not fully accepted, largely because his work tended to break established theory.

Fresnel, a confirmed Royalist, was removed from his position when Napoleon returned to power, being reappointed when Napoleon was removed a second time. The sickly young man continued to work on highways and bridges, but experi-

An unrecognized genius in his own country, Augustin-Jean Fresnel designed one of the great inventions in the history of lighthouses—the lens named after him. (Smithsonian Institution, courtesy U.S. Lighthouse Society)

mented with light in his spare time. Necessity caused his experiments to take on an almost Rube Goldberg quality.

While working on roads in Brittany, Fresnel's only sources for experiments with light were the sun, a micrometer fashioned from wires and cardboard, which was later replaced by a slightly better instrument constructed by a village blacksmith. For an experiment on a short focus lens, Fresnel used a piece of paper with a small hole for viewing over which he placed a drop of honey that took the place of a lens. The amazing fact is that his experiments, even using these jury-rigged instruments, proved successful. Twice he was awarded a few months off from the roads to travel to Paris to work with real instruments. He presented more papers and some recognized Fresnel's genius, but always he was returned to the roads. At one point, he was even sent to Rennes and placed in charge of a workhouse for the unemployed.[72]

Around 1821, Fresnel, in failing health, was temporarily appointed secretary to the Commission for Lighthouses. During this period, he managed to have his lens installed in the principal lighthouses along the French coast. The French lights became so superior that soon lighthouses all over Europe were being fitted with the lens. The use of the Fresnel lens in the United States, as already mentioned, was delayed by Pleasonton but, after lighthouses passed from his control, the lens was quickly adopted.

FRESNEL ORDERS

THE ORDERS

THE DIAGRAM HERE SHOWS THE RELATIVE SIZES OF 1ST THROUGH 6TH ORDER LENSES SUPERIMPOSED, WITH THE EXCEPTION OF THE 3½ ORDER SIZE WHICH WAS DEVELOPED AFTER THE ORIGINAL ORDERS WERE STANDARDIZED. FRESNEL ESTABLISHED FOUR ORDERS OF LIGHTS, BASED UPON THE NUMBER OF CONCENTRIC CYLINDRICAL WICKS IN THEIR LAMPS—THE 1ST ORDER HAVING FOUR WICKS, THE 4TH ONE WICK. THE 3RD AND 4TH FRESNEL ORDERS EACH HAD TWO SIZES: FRESNEL'S ORIGINAL 3RD AND 4TH ORDER LARGE SIZES CORRESPOND TO THE PRESENT 3RD AND 5TH ORDER LENSES, AND THE ORIGINAL 3RD AND 4TH ORDER SMALL SIZES ARE NOW KNOWN AS THE 4TH AND 6TH ORDER LENSES IN THE UNITED STATES.

DEVELOPMENT OF UPPER AND LOWER REFLECTING PANELS

WHEN FRESNEL FIRST DEVELOPED HIS LENS DESIGN IN 1821, HE RELIED UPON METALLIC REFLECTORS TO CAPTURE AND REDIRECT LIGHT ESCAPING ABOVE AND BELOW THE LENS PANELS. AFTER HIS DEATH IN 1857, BANKS OF REFLECTING GLASS PRISMS WERE SUBSTITUTED FOR THE REFLECTORS, BECAUSE THEY ABSORBED LESS LIGHT AND REQUIRED LESS MAINTENANCE. ULTIMATELY, A COMPLETE GLASS LENS WITH CENTRAL "DRUM" AND UPPER AND LOWER PRISMS TRANSMITTED 5/6 OF THE AVAILABLE LAMP LIGHT, A CONSIDERABLE IMPROVEMENT OVER THE 1/6 EMISSION CAPABILITIES OF PRE-1821 METALLIC REFLECTOR LIGHTS. AS AN ADDITIONAL IMPROVEMENT (ALONG WITH GREATER VISIBILITY OF THE LIGHT AS SEEN FROM THE SEA), THE LAMPS CONSTRUCTED FOR THESE LENSES CONSUMED LESS FUEL (OIL).

LENS MANUFACTURERS

SEVERAL FRENCH COMPANIES IN THE VICINITY OF PARIS WERE RESPONSIBLE FOR ALL THE FRESNEL LENSES USED IN THE UNITED STATES: LETORNEAU & CO., HENRY-LEPAUTE, L. SAUTTER & CO. (LATER SAUTTER, LEMONER & CO.), AND BARBIER & FENESTRE (LATER BARBIER, BENARD, TURENNE). MANY OF THEIR LENSES REMAINED IN USE FOR OVER 80 TO 100 YEARS.

LAMPS AND LAMP FUELS

MID-EIGHTEENTH CENTURY EUROPEAN LIGHTHOUSES USED PRIMARILY COLZA AND RAPESEED OILS IN THEIR LAMPS, WHILE THE UNITED STATES RELIED HEAVILY ON SPERM WHALE OIL UNTIL SPERM WHALES BECAME SCARCE. AS LAMP TECHNOLOGY PROGRESSED, KEROSENE LAMPS AND THEN THE INCANDESCENT OIL VAPOR (IOV) LAMPS WERE DEVELOPED, FOLLOWED BY VARIOUS KINDS OF ELECTRIC LAMPS, SUCH AS THE 1000 WATT 120 VOLT QUARTZ-IODIDE LAMP OFTEN USED TODAY.

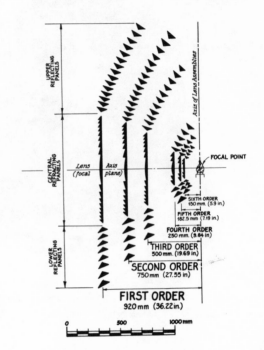

NOTES

1) LENS SECTIONS SHOWN ARE BASED ON PLATE 3, FIGS. 1-6, FROM MEMOIR UPON THE ILLUMINATION AND BEACONAGE OF THE COASTS OF FRANCE BY M. LEONCE REYNAUD (WASHINGTON, D.C.: GOVERMENT PRINTING OFFICE, 1876).

2) LENS SECTIONS SHOWN ABOVE ARE SUPERIMPOSED AT THE SAME SCALE FOR COMPARATIVE PURPOSES ONLY. NO LENS WAS EVER CONSTRUCTED WITH ALL ORDERS COMBINED.

3) 3½ (375mm) AND HYPERRADIAL (1330mm) ORDERS NOT SHOWN (NOT EXTANT IN 1873).

COMPARATIVE TABLE OF LENS ORDERS

ORIGINAL FRESNEL ORDERS		N° OF LAMP WICKS	OIL CONSUMPTION PER HOUR		RELATIVE BRIGHTNESS (6TH ORDER=1)	MODERN LENS ORDERS (U.S.A.)	FOCAL LENGTH		HEIGHT OF COMPONENTS								APPROXIMATE WEIGHT OF ASSEMBLED LENS		USE
ORDER			gm.	oz.			mm.	in.	LOWER REFLECTOR		CENTRAL REFRACTOR		UPPER REFLECTOR		TOTAL		kg.	lbs.	
									mm.	in.	mm.	in.	mm.	in.	mm.	in.			
1ST		4	750	26.25	17.69	1ST	920	36.22	539	21.22	980	38.58	1001	39.40	2590	101.97	5800	12800	LARGEST SEACOAST LIGHTS
2ND		3	500	17.5	11.54	2ND	750	27.55	378	14.88	854	33.62	810	31.89	2069	81.46	1600	3,530	GREAT LAKES LIGHTHOUSES, SEACOASTS ISLANDS, SOUNDS
3RD	LARGE	2	200	7	3.85	3RD	500	19.69	278	10.94	660	25.98	593	23.35	1576	62.05	900	1,985	SEACOAST, SOUNDS RIVER ENTRY, BAYS, CHANNELS, RANGE LIGHTS
	SMALL	1	150	5.25	2.31	4TH	250	9.84	144	5.67	300	11.81	358	14.09	722	28.43	200 TO 300	440 TO 640	SHOALS, REEFS, HARBOR LIGHTS, ISLANDS IN RIVERS AND HARBORS
4TH	LARGE	2	90	3.15	1.23	5TH	1825	7.19	105	4.13	226	8.90	196	7.72	541	21.30	120 TO 200	265 TO 440	BREAKWATERS RIVER LIGHTS, CHANNEL, SMALL ISLANDS IN SOUNDS.
	SMALL	1	90	3.15	1	6TH	150	5.9	84	3.31	180	7.09	157	6.18	433	17.05	100	220	PIER OR BREAKWATER LIGHTS IN HARBORS.

NOTES TO TABLE

1. TOTAL HEIGHT INCLUDES INTERIOR METAL FRAMES.

2. COMPARATIVE BRIGHTNESS OF LIGHTS IS BASED ON A TABLE IN M. LEONCE REYNAUD, MEMOIR UPON THE ILLUMINATION AND BEACONAGE OF THE COASTS OF FRANCE (WASHINGTON, D.C.: GOVERNMENT PRINTING OFFICE, 1876), P. 55.

3. RANGES OF LIGHTS ARE NOT GIVEN AS THIS IS PARTIALLY DEPENDENT ON HEIGHT OF THE LIGHT ABOVE SEA LEVEL.

4. WEIGHTS OF LENSES AND OIL CONSUMPTION RATES ARE BASED ON PHARES ET FANAUX

LENTICULAIRES DESCRIPTION ET PRIX DE APPAREILS CONSTRUITS PAR L. SAUTTER ET CIE (PARIS: IMPRIMERIE CENTRALE DE NAPOLEON CHAIX ET CIE, 1858)

5. ALL OPTICAL DESIGN AND PRODUCTION IS CONDUCTED IN THE METRIC SYSTEM WORLDWIDE, HENCE METRIC DIMENSIONS

ARE GIVEN AS PRIMARY FIGURES, ENGLISH CONVERSIONS AS SECONDARY.

6. 3½ (375 mm) AND HYPERRADIAL (1330mm) ORDERS NOT SHOWN (NOT EXTANT IN 1873).

DELINEATED BY: MABEL A. BAIGES, 1988			
SOUTHEAST LIGHT RECORDING PROJECT HISTORIC AMERICAN ENGINEERING RECORD NATIONAL PARK SERVICE UNITED STATES DEPARTMENT OF THE INTERIOR	BLOCK ISLAND SOUTHEAST LIGHT - 1874 SPRING STREET AND MOHEGAN TRAIL, AT MOHEGAN BLUFFS, LIGHTHOUSE COVE	SHEET 8 of 12	HISTORIC AMERICAN ENGINEERING RECORD
	BLOCK ISLAND WASHINGTON COUNTY RHODE ISLAND		RI-27

IF REPRODUCED, PLEASE CREDIT: HISTORIC AMERICAN ENGINEERING RECORD, NATIONAL PARK SERVICE, NAME OF DELINEATOR, DATE OF THE DRAWING

These were the orders and descriptions of Augustin-Jean Fresnel's legacy to aids to navigation.

SIMPLE PLANO-CONVEX LENS

A PLANO-CONVEX LENS HAS ONE FLAT FACE AND ONE CONVEX FACE WHOSE SURFACE IS A SPHERICAL SEGMENT. LENSES TAKE ADVANTAGE OF THE FACT THAT OPTICALLY TRANSPARENT, DENSER-THAN-AIR MATERIALS BEND, OR REFRACT, LIGHT RAYS. BY THEIR GEOMETRY, PLANAR CONVEX LENSES CAN BEND LIGHT EMITTED FROM A POINT SOURCE ON ONE SIDE OF THE LENS IN SUCH A MANNER THAT A BEAM IS CREATED PARALLEL TO THE LENS AXIS ON THE OTHER SIDE.

EIGHTEENTH-CENTURY LIGHTHOUSE OPTICS DEPENDED ON PARABOLIC METALLIC REFLECTORS TO COLLECT AND FOCUS LAMPLIGHT INTO BEAMS THAT WOULD HAVE ENOUGH INTENSITY TO BE SEEN BY SHIPS. THESE REFLECTORS REQUIRED FREQUENT CLEANING AND POLISHING, AND WERE INEFFICIENT TO THE EXTENT THAT THEY TRANSMITTED EFFECTIVELY ONLY 1/6 OF THE LIGHT PRODUCED BY A LAMP.

SINGLE GLASS LENSES OF GREAT SIZE WERE INSTALLED IN SEVERAL MID-EIGHTEENTH CENTURY LIGHTHOUSES IN ENGLAND AND IRELAND, BUT THE SIZE AND THICKNESS OF THE GLASS ABSORBED TOO MUCH OF THE AVAILABLE LIGHT. IN ADDITION, SUCH LENSES WERE HEAVY, THUS DIFFICULT TO MANUFACTURE, INSTALL, AND MAINTAIN.

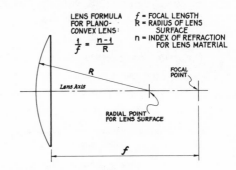

LENS FORMULA FOR PLANO-CONVEX LENS:

$$\frac{1}{f} = \frac{n-1}{R}$$

f = FOCAL LENGTH
R = RADIUS OF LENS SURFACE
n = INDEX OF REFRACTION FOR LENS MATERIAL

BUFFON LENS (ca. 1748)

AS EARLY AS 1748, GEORGE LOUIS LECLERC, COMTE DE BUFFON, HYPOTHESIZED THAT THE EXCESSIVE WEIGHT AND LIGHT ABSORPTION OF LARGE GLASS LENSES COULD BE GREATLY REDUCED BY GRINDING A CAST SLAB OF GLASS (1220 mm (48") DIAMETER BY 65 mm (2½") THICK) INTO A SHAPE AS SHOWN AT RIGHT. THE CURVED SURFACES OF BUFFON'S HYPOTHETICAL LENS RESULTED WHEN THE SURFACE OF A SIMPLE SPHERICAL LENS WAS COLLAPSED IN A SERIES OF CONCENTRIC STEPS. MARIE J.A.N. DE CARITAT, MARQUIS DE CONDORCET, SUGGESTED IN 1773 THAT SUCH A LENS WOULD BE EASIER TO MANUFACTURE IN SEVERAL PIECES, RATHER THAN FROM ONE LARGE LENS. NEITHER MAN, HOWEVER, CONSIDERED EMPLOYING SUCH A LENS IN LIGHTHOUSES, NOR DID THEY DEVELOP CALCULATIONS THAT WOULD OPTIMIZE THE PERFORMANCE OF SUCH LENSES.

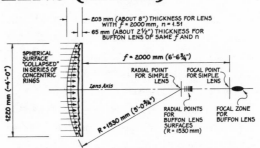

203 mm (ABOUT 8") THICKNESS FOR LENS WITH f = 2000 mm, n = 1.51
65 mm (ABOUT 2½") THICKNESS FOR BUFFON LENS OF SAME f AND n

SPHERICAL SURFACE "COLLAPSED" IN SERIES OF CONCENTRIC RINGS

f = 2000 mm (6'-6¾")

RADIAL POINT FOR SIMPLE LENS

FOCAL POINT FOR SIMPLE LENS

1220 mm (-4'-0")

R = 1530 mm (5'-0¾")

RADIAL POINTS FOR BUFFON LENS SURFACES (R = 1530 mm)

FOCAL ZONE FOR BUFFON LENS

FRESNEL LENS (1821)

UNAWARE OF THE WORK OF BUFFON AND CONDORCET, IN 1821 AUGUSTIN-JEAN FRESNEL (FREH-NEL) DESIGNED EIGHT THEORETICALLY IDENTICAL LENSES COMPOSED OF INDIVIDUALLY MANUFACTURED ANNULAR PRISMS FOR USE IN AN EXPERIMENTAL LIGHT AT THE RENOWNED CORDOUAN LIGHTHOUSE IN FRANCE. UNLIKE THE LENSES OF HIS PREDECESSORS, FRESNEL'S LENS INCLUDED CORRECTIONS BASED ON FORMULAS HE HAD DEVELOPED FOR SPHERICAL ABERRATIONS, LIGHT WAVE INTERFERENCE, DOUBLE-REFRACTION AND POLARIZATION. IN SECTION, EACH ANNULAR PRISM HAS A UNIQUE RADIAL POINT WHICH LIES OFF THE LENS AXIS. DUE TO THE SYMMETRY OF A FRESNEL LENS, THE DIAGRAM AT RIGHT SHOWS RADIAL POINTS FOR ONLY HALF OF A LENS. THE METRIC DIMENSIONS GIVEN ARE FOR A "FIRST-ORDER" LENS, SUCH AS THE LENS APPARATUS IN USE AT THE BLOCK ISLAND SOUTHEAST LIGHT IN 1988. THE PLANAR CONVEX LENS FORM WAS CHOSEN BY FRESNEL BECAUSE ITS FLAT BACK PERMITTED EASY ASSEMBLY OF LENS PIECES ON A FLAT TABLE, WHERE THEY WERE GLUED TOGETHER AND MOUNTED IN BRONZE FRAMES.

THE CORDOUAN LENSES WERE GROUND BY A PARISIAN OPTICIAN, JEAN-BAPTISTE FRANCOIS, FROM "CROWN" GLASS PRODUCED BY THE ROYAL GLASSWORKS AT ST. GOBAIN NEAR PARIS. CROWN GLASS POSSESSES SEVERAL ADVANTAGES OVER "FLINT" GLASS, WHICH COULD ALSO HAVE BEEN USED FOR LENSES. ALTHOUGH CROWN GLASS HAS A GREENISH TINT AND A LOWER INDEX OF REFRACTION THAN FLINT GLASS, CROWN GLASS IS HARDER, LESS DENSE, AND LESS REACTIVE WITH ATMOSPHERIC CONTAMINANTS THAN LEAD-BEARING FLINT GLASS. LIGHTHOUSE LENSES HAVE EMPLOYED CROWN GLASS ALMOST EXCLUSIVELY SINCE THE 1820's.

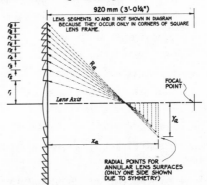

920 mm (3'-0¼")

LENS SEGMENTS 10 AND 11 NOT SHOWN IN DIAGRAM BECAUSE THEY OCCUR ONLY IN CORNERS OF SQUARE LENS FRAME.

FOCAL POINT

Lens Axis

RADIAL POINTS FOR ANNULAR LENS SURFACES (ONLY ONE SIDE SHOWN DUE TO SYMMETRY)

Lens Segment	a=1	2	3	4	5	6	7	8	9	10	11
x_a	454.79	488.55	513.38	540.71	563.27	588.00	614.35	636.90	660.11	683.41	712.79
y_a	00.00	13.08	31.72	57.00	84.86	114.93	151.50	189.55	230.17	280.60	323.00
r_a	140	208.15	262.40	309.20	350.50	387.44	422.23	456.23	490.00	523.33	555.35
R_a	483.50	543.60	598.62	659.77	719.84	779.48	846.45	911.30	980.50	1057.70	1136.01

ALL DIMENSIONS IN MILLIMETERS (DIVIDE BY 25.4 TO CONVERT TO INCHES). DATA BASED ON PLATE XII IN ALAN STEVENSON, ACCOUNT OF THE SKERRYVORE LIGHTHOUSE, WITH NOTES ON THE ILLUMINATION OF LIGHTHOUSES, EDINBURGH, SCOTLAND: ADAM AND CHARLES BLACK, 1848.

DELINEATED BY MABEL A. BAIGES, 1988

SOUTHEAST LIGHT RECORDING PROJECT
NATIONAL PARK SERVICE
UNITED STATES DEPARTMENT OF THE INTERIOR

BLOCK ISLAND

BLOCK ISLAND SOUTHEAST LIGHT - 1874
SPRING STREET AND MOHEGAN TRAIL AT MOHEGAN BLUFFS, LIGHTHOUSE COVE
WASHINGTON COUNTY

RHODE ISLAND

SHEET 7 OF 12

HISTORIC AMERICAN ENGINEERING RECORD
RI-27

IF REPRODUCED, PLEASE CREDIT: HISTORIC AMERICAN ENGINEERING RECORD, NATIONAL PARK SERVICE, NAME OF DELINEATOR, DATE OF THE DRAWING

(National Park Service)

Some of the Fresnel lenses used in this country were huge. The light at Makapuu Point, Hawaii, for example, weighed nine tons. But the lighthouse at Navesink, New Jersey, installed in 1898, with an electric arc lamp could boast an incredible sixty million candlepower. The size of the lenses may be better understood by perusing the illustration on pages 24 and 25, which gives the sizes of the various lenses.

Augustin-Jean Fresnel died on Bastille Day 1827, at the age of thirty-nine. Most of his short life and genius was wasted on the roads of France. The value of Fresnel's lens is incalculable. The lens is just one small part of this genius's work; he was virtually ignored during his lifetime. Most of Fresnel's work did not become accepted until generations after his death, with some of that long-ago work still being found today in physics textbooks.

Fresnel remains a mystery. One biographer wrote: "Who was he? He is hard to get at." Georges A. Boutry wrote that while visiting with the mayor of Ville d'Avray, the official went on for great lengths about the famous personalities who had graced the small village. At last the mayor paused for breath, and Boutry asked: "'And what of Fresnel?'" "'Who was he?' said the Mayor."[73] "There is much more I should have done," Fresnel once said. Genius to the end, he remarked, "All the compliments that I have received never gave me so much pleasure as the discovery of a theoretic truth, or the confirmation of a calculation by experiment."[74]

More than a century and a half after the Fresnel lens was introduced into lighthouses, many of the lenses are still in use today. Once the Fresnel lens was in operation, there was no need for further work on developing lenses. Technology could concentrate on better sources for the light.

VARIOUS TYPES OF LAMPS were used in United States lighthouses. Carcel, Lepaute, and Funck were the names of a few of the lamps. The Funck was widely used, perhaps influenced by the fact that Funck was employed at the service's principal supply depot. All the lamps used a concentric wick. The only difference lay in the method used to get the oil to the wick.[75]

As mentioned earlier, the first illuminates for lighthouses were wood and coal. In the United States, whale oil was the primary fuel for many years. There were two grades: a thick one, called summer oil, and a thin winter oil. In colder climates, however, even the thin oil would thicken in winter and a stove had to be kept in the lantern room of the lighthouse to liquefy the oil.[76]

Eventually, only sperm oil was used in the lighthouses. This type of oil was considered to be of the best quality and would produce the better light. By around the middle of the nineteenth century, however, as whalers nearly depleted the whale population, the price of the oil rose steadily. In 1840–1841, the price of a gallon of sperm oil was fifty-five cents and rocketed to $2.25 per gallon by 1855. The new lighthouse board would have to begin the search for a cheaper source of fuel.[77]

At the end of 1851, the United States, despite its need for maritime trade, had a second-rate system of lighthouses. The major blame lay with Stephen Pleasonton, the fifth auditor. Pleasonton (as was previously stated) was a hard-working bureau-crat who had no knowledge of maritime affairs, but knew how to keep a tight lid on a budget, being extremely proud that "no bureau of the federal government turned back to the general fund more money than his did."[78] He was not disabused of this practice by a penny-pinching Congress. By 1851, the increasing number of complaints about lighthouses finally forced Congress to act. When Pleasonton was relieved of his duties and control of the lighthouses passed in October 1852 to the new U.S. Lighthouse Board, a new era of aids to navigation was ushered into the United States.

A Brightly
Burning Light

THE ESTABLISHMENT OF THE nine member U.S. Lighthouse Board brought stability to the lights. The eminent lighthouse historian, F. Ross Holland, has noted that during its tenure the "aids to navigation in the United States improved dramatically."[1] The board accomplished this largely by making sure that the mariner could depend upon a good warning light. The members of the board had a wide range of experience and knowledge; they were able to translate this into the ability of seeking good keepers and insuring that the illuminants were the best that could be provided. The board also sought constant improvements in equipment. They helped maintain the level of good keepers and, most important, published good instructions to those who kept the lights. In short, the board "raised the reputation of the United States' lighthouses from the bottom of the heap to the top. . . ."[2]

The board reorganized the system by dividing the country into twelve lighthouse districts. The First District began in Maine and the twelfth took in the west coast of the United States. The board established an inspector for each district. This official was either a navy or an army officer and was responsible for the maintenance of the lights and the discipline of the keepers. Eventually, the board felt there needed to be further division of duties and created a district engineer. At that time, the inspector was generally a naval officer who acted as a district superintendent and carried out administrative duties in addition to checking on the abilities of keepers. The engineer tended to be an army officer and handled the building of "lighthouses, . . . keeping them in repair, and with the purchase, the setting up, and the repairs of the illuminating apparatus."[3]

The board felt there should be a central source of supplies and established a depot on Staten Island. As the years passed and the service expanded, the board situated depots in each district. The Staten Island unit, however, shipped needed supplies to the district depots. Staten Island was charged with the testing of oil used in lighthouses.[4]

The lighthouse board began to establish a clear set of rules and procedures and, most important, they published the new instructions for lighthouse keepers. Every aspect of managing a lighthouse was set down. At stations with two keepers and only one lamp room, for example, the instructions clearly stated that "the daily duty shall be laid out in two departments, and the lightkeepers shall change from one department to the other every Sunday night."[5] The various pieces of equipment were described, along with how to maintain them correctly. By 1902, the instructions had grown to a book of fifty-five pages, along with many plates of photographs showing equipment and many detailed drawings. The instructions now made it almost impossible not to keep a good light. Furthermore, inspectors now had a standard upon which to judge keepers. The U.S. Lighthouse Board was usually not slow in removing an inept keeper.

One of the most important publications that the U.S. Lighthouse Board established was an annual *Light List*. The publication recorded all the aids to navigation in the United States, describing their locations and characteristics. The board also established a system of correcting the list, known as the Notice to Mariners, which let navigators know of changes and additions between the publications of the *Light List*.

The U.S. Lighthouse Board remained under the Treasury Department and, as mentioned earlier, was organized with the secretary of the treasury as the ex officio president. There was also a chairman of the board. The first to hold this position was Como. William B. Shubrick, of the U.S. Navy. The board then appointed five standing committees: finance, engineering, light vessels, lighting, and experiments.

A little more than a decade after being established, the U.S. Lighthouse Board's continued efforts to improve the aids to navigation were thwarted by the Civil War. The war would have priority over any construction and lights. A second light at Southwest Pass, Louisiana, for example, had to be put on hold until after the war. Other lights were destroyed by either Confederate or Union forces. The Fresnel lens at Tybee Island, Georgia, was removed and the tower put to the torch. The lighthouses of the Gulf of Mexico region began to feel the war in April 1861 when the lighthouse supply vessel *Guthrie* arrived in Galveston, Texas, with supplies for the lights of the district. Prior to that, in February 1861, the district inspector, Lt. Joseph Fry, U.S. Navy, resigned because Louisiana had seceded. Interestingly, six months after the war ended, Fry applied for a position with the lighthouse service to care for the buoys in the Eighth and Ninth Lighthouse Districts. Not too surprisingly, passions concerning the war were still high and the board replied that it could not "consent to the employment of Mr. Fry in any capacity under the Light House Establishment."[6] The Confederates disabled many lights along the coast, for they knew the lights would be of more use to the powerful federal navy. The Union forces returned many of the sites as soon as possible. For example, the Confederates chose the site of the light at Barataria Bay, Louisiana, for a fort to protect New Orleans. Before completing the fortification, Union forces captured the site in 1862 and asked the U.S. Lighthouse Board to reestablish the light.[7] After the end of hos-

tilities, the U.S. Lighthouse Board continued its work on improving the aids to navigation in this country. The next major change for aids to navigation would not take
place until the first decade of the twentieth century.

From its inception, the U.S. Lighthouse Board had been dominated by military
officers. It will be recalled that the first board consisted of three naval officers and
four army officers. Indeed, the act that created the board specifically stated that the
board would be made up of seven members who were either from the army or the
navy. Furthermore, many of the district inspectors and engineers were from the military. By the twentieth century, Congress wished to have the lighthouses under civilian control, feeling it had "moved far from the purpose of its founders."[8] At the same
time, decision makers felt the board was becoming too cumbersome an administrative unit to effectively manage the large numbers of lights. For example, in the
1850s, when the U.S. Lighthouse Board was established, there were 297 major
lights in the United States. By 1910, the number had ballooned to 1,397.[9]

In 1910, Congress abolished the U.S. Lighthouse Board and created the Bureau
of Lighthouses. Seven years previous to this decision, in 1903, Congress had created the Department of Commerce and Labor and transferred the lights from
Treasury to the new department. Congress opted to have the new Bureau of
Lighthouses remain in Commerce. Even though one intent of the change was to
limit military officers from serving with the Bureau of Lighthouses, the act did give
the president the power to detail officers from the army engineers as consultants or

*Maj. Hartman Bach, a district inspector for the West Coast, believed the lighthouse archives should
contain pictures and plans of all lighthouses. Many of his sketches were rendered in watercolor and
are the best surviving contemporary views of the early lighthouses of the West Coast. This is Bach's
depiction of California's Point Bonita Light Station around 1856.*

Bach's rendering of Point Loma Light Station in California, where the light was first lit in 1855.

supervisors in construction.[10] Holland has rightfully pointed out that the fine reputation of the lights in this country was "accomplished because of, rather than in spite of, the military dominance of the lighthouse service. . . . The country owes a tremendous debt of gratitude to the Lighthouse Board and its military members."[11] In fact, military men contributed to some of the more amazing engineering feats undertaken and to technically improving the lights. Civil War buffs, for example, may be surprised to learn that George G. Meade (the Civil War commander of the Union forces at the decisive battle of Gettysburg) while a lieutenant in the army engineers devised a lamp during his assignment to lighthouses. The planning and construction of the major engineering feat required to establish the Spectacle Reef light in Lake Huron was accomplished by then Maj. O. M. Poe, who would become Gen. William B. Sherman's chief engineer during the march to the sea.[12]

Congress dictated there would be a commissioner to head the bureau with a deputy, a chief constructing engineer, and a superintendent of naval construction as his chief officers. The act also started a reorganization of the lighthouse service, and allowed the new commissioner to increase the number of districts to no more than nineteen; each district was to be headed by an inspector, who, two years after passage of the act, was a civilian. By August 1912, civilians headed all but three of the districts, and these inspectors were for the most part career lighthouse service employees. The commissioner of lighthouses, however, gave a broad interpretation to the portion of the act that allowed the president to detail army engineers for consulting and supervising construction. An officer from the Corps of Engineers was assigned to each district. The officer had virtually no role in the district's administrative command structure, although his official title was superintendent. He concerned himself almost entirely with construction and repair work in the district and reported to the civilian district inspector.

The first commissioner of lighthouses, and the fifth pivotal person in the history of aids to navigation in this country, was George R. Putnam, who had no military background. Putnam was born in Davenport, Iowa, in 1865, and led an adventurous life. As the son of an attorney, he began the study of law but soon decided to enter engineering. He attended Rose Polytechnic Institute in Indiana. Upon graduation, he joined the Coast and Geodetic Survey and for the next twenty years was involved in map making. Until the turn of the century, most of his assignments took place in polar or subpolar regions. He was on the survey to Alaska that established the southeast boundary of that territory. He went to Greenland with the American Arctic explorer Robert E. Peary in 1896 to retrieve a large meteorite that had fallen near Cape York. Putnam's duties were to undertake magnetic and gravity measurements. The next year saw him assigned to survey the Pribilof Islands in the Bering Sea, and he spent two more years in surveying and mapping the mouth of the Yukon River. In 1900, Putnam's duties took him to a much warmer climate when he was appointed director of coast surveys in the Philippines. The following six years were spent supervising and charting the coasts of these islands. While in the Philippines, Putnam met William Howard Taft, who was governor of the islands. His last assignment in the Coast and Geodetic Survey was a four-year stint at headquarters in Washington, D.C. While in Washington, Secretary of Commerce and Labor Charles Nagel asked Putnam to head the newly created bureau. After gaining a "fairly reasonable assurance of freedom from political interference," Putnam became the new commissioner of lighthouses. From all indications, Putnam was successful in keeping the service free from political patronage, as shown by his choice of inspectors. The commissioner appointed career government employees, most of whom were veterans of the lighthouse service, to these important positions. Ralph H. Goddard, for example, began as a fireman aboard one of the service's vessels in 1887. Later he became a ship's assistant engineer and, in 1904, he came ashore as a supply clerk in the Third District. Two years later, Putnam appointed him inspector of the Second District.[13]

By all accounts, Putnam was masterful at handling Congress. The legislature, in turn, seemed to respect the commissioner. His organization also greatly benefited from Putnam's skill as an excellent writer.

The U.S. Lighthouse Board continued to search for a good, but cheap, illuminant for the lamps of the lighthouse service. When the price of whale oil began to rise, the board began to experiment in earnest with other fuels. One of these, colza oil, which is derived largely from wild cabbage seeds, had all the necessary properties of a cheap fuel source, except it had to be imported. The board tried to interest U.S. farmers

Harriet E. Colfax struggled with the problems of being the head keeper of the Michigan City Light Station in Indiana from 1861 to 1904.

in growing this crop. Farmers failed to respond in large enough numbers to continue with this source.[14]

Joseph Henry, of the Smithsonian Institution and head of the board's committee on experiments, had previously tried lard oil, but this proved unsuccessful. Not altogether discouraged that the product was unsuitable for the service, Henry began to resume testing. His renewed work revealed that if the lard oil was heated to a high enough temperature it burned well. This oil would be cheap and in great supply, so the U.S. Lighthouse Board quickly adopted its use. By 1867, lard oil was being used exclusively in the larger lamps. One problem with the oil, however, was that it had to be at a certain temperature to be useful. Witnesses recall the problems that Harriet Colfax, the diminutive keeper of the light station at Michigan City, Indiana, had in cold weather. One of the duties of Keeper Colfax was maintaining a post light on a pier, which forced her to cross a creek by small boat, walk a distance along the creek, climb an elevated walkway to reach the light and, finally, climb the ladder to the light. One stormy night, Colfax heated the lard oil and started off on her journey. High, gusting winds and waves lashed at her. Twice driven back, she eventually made it to the lantern, but the oil had congealed. She had to retreat to the keeper's quarters to reheat the oil. In the process she fell twice. Undaunted, she returned to the stove, reheated the oil, and rejoined the fight against the elements. This time, although bruised and soaked, she was successful. (The author can remember when five U.S. Coast Guardsmen had to struggle along the elevated walkway in a raging

January snowstorm to relight the Michigan City east pierhead light. Like Harriet Colfax we made two trips to the light. Unlike Colfax, we had foul weather gear and life jackets and the assurance of each other to help us along. All of this makes Harriet Colfax's feat all the more remarkable.[15])

By the 1870s, the U.S. Lighthouse Board was once again experimenting with illuminants. This time the committee on experiments investigated what was then known as mineral oil, now called kerosene. Kerosene at first was considered too dangerous for use in lighthouses. A keeper on Lake Michigan in 1864, however, took matters into his own hands and used the oil. A series of explosions shook the light tower, blew the lantern from the tower and wrecked the lens. This put a stop to the "dangerous" oil. Putnam noted that the U.S. Lighthouse Board hesitated in 1875 "to endanger the lives of employees and valuable property" with the oil but, within two years, kerosene became the illuminant in the smaller lights. The service converted all other order lenses to kerosene as new lens lamps became available and, by 1885, most lights used mineral oil.[16]

The incandescent oil vapor (IOV) lamp was the next major change in the lights. North Hook beacon, Sandy Hook, New Jersey, received the first such lamp. In this lamp, kerosene is forced into a vaporizer chamber where it strikes the hot walls and is changed into a vapor. The gas goes through a series of small holes to a mantle where it burns like a glowing gas ball. Many will recognize this description as the operating principle behind some camp lanterns. This device increased the candle power and brilliance many fold, without an increase in fuel consumption. At the light station at Cape Hatteras, North Carolina, for example, the power of the light increased from 27,000 to 80,000 candle power, while the oil consumption went from 2,280 to 1,300 gallons a year. This was the last step in the refinement of flame for lighthouse illumination.[17]

The first use of electricity in lighthouses was at Dungeness Lighthouse on the coast of England in 1862, but the experiment was abandoned. In the United States, electricity for lighthouse purposes began with the placing of an arc light in the Statue of Liberty in 1886. The conversion went slowly because most lights, by their very nature, were located some distance from the early power lines. In some areas, generators produced electricity. During the 1920s and 1930s, the Bureau of Lighthouses converted most lights to electricity. One wonders if anyone recognized at the time that this last improvement in the illumination of lights spelled the beginning of the end for a way of life.[18]

The U.S. Lighthouse Board experimented over the years with the use of coal and natural gas. At the light station at Jones Point, near Washington, D.C., the board installed pipes that ran from the Alexandria, Virginia, gas works to the light. The pipes proved a constant source of problems because of erosion and water and, in 1900, the board had the light converted to oil. In general, the U.S. Lighthouse Board felt that "gas was simply not satisfactory as an illuminant, and consequently, gas never really got past the experimental stage in American lighthouses."[19]

California's St. George Reef Light Station took ten years to build because of its exposed and treacherous location.

While the coming of electricity spelled great changes for the lighthouse service, another momentous change began a scant two years after the establishment of the Bureau of Lighthouses. Again, one wonders if anyone had enough foresight to recognize the portent of things to come. President William Taft presided over a nation that was in the throes of political and social reform, which would become known as the Progressive Movement. Taft had noted that the huge growth of the United States came through the "use of the principle of organization and combination and through the development of our natural resources." He went on to note how "we nearly transferred complete political power to those who controlled corporate wealth and we were in danger of a plutocracy." This concern over a plutocracy was one of the reasons for the Progressive Movement and the call for reform in both social and political matters.[20]

Quite naturally the national government came under scrutiny. Following his views

If a light station rested on a high promontory, the tower could be short. At Point Reyes, California, the iron tower is 23 feet tall at the focal plane and sits 296 feet above sea level. Some of the first locations chosen for West Coast lighthouses sat so high that the lights could not be seen by ships in low-lying fog.

of "organization and combination," Taft, under acts of 25 June 1910 and 3 March 1911, appointed a commission to find ways to improve the economy and efficiency of the federal government. Frederick A. Cleveland, an economist and financial advisor to the president, headed what was to become known as the Taft Commission. The small maritime police force and safety forces of the federal government were some of the areas to come under close examination.

Within the Treasury Department, the U.S. Revenue Cutter Service and the U.S. Life-Saving Service received special attention.[21] The commission eventually recommended that the U.S. Life-Saving Service be discontinued and its stations be made a part of the Bureau of Lighthouses, while the U.S. Revenue Cutter Service should be dissolved and its duties and cutters spread throughout the federal government. When asked about his views of the proposal, Charles Nagel, Secretary of Commerce and Labor, quite naturally was agreeable to the changes, but still offered as an alternative to the commission's report a plan to create a federal maritime safety and law enforcement agency made up of the Bureau of Lighthouses, the U.S. Life-Saving Service, and the U.S. Revenue Cutter Service. Nagel's plan did not carry. In last minute maneuverings, the U.S. Revenue Cutter Service and U.S. Life-Saving Service were amalgamated on 20 January 1915, to form the U.S. Coast Guard.[22]

George R. Putnam held sway over the Bureau of Lighthouses for twenty-five years until his retirement in 1935. Under his stewardship, aids to navigation in the United States continued as the best in the world. By 1924, the U.S. Lighthouse Service was the largest such agency in the world with 16,888 aids to navigation. Interestingly, while the aids to navigation had more than doubled, the number of employees had decreased by nearly 20 percent. The reason for the decrease in personnel was due in large part to the adoption of technological advances, such as the use of electricity, that reduced the number of people needed to attend the lights. Technology would have great consequences for the future of lighthouses in the United States.[23]

Putnam oversaw the introduction of radio beacons at lighthouses, the use of the electric buoy, and the application of electricity to fog signals. One of his greatest accomplishments for the people of the Bureau of Lighthouses was a retirement system for field employees and those serving on lighthouses.[24] Putnam also produced one of the two best books to date on the lighthouse service as a national entity.[25]

Less than five years after Putnam's retirement, Secretary Nagel's rejected plan of placing the U.S. Revenue Cutter Service, the U.S. Life-Saving Service, and the lighthouses under one agency came into being. The U.S. Lighthouse Service was amalgamated into the U.S. Coast Guard.

Citing "the interest of economy and efficiency," President Franklin D. Roosevelt announced his Reorganization Plan II. Part of the language of this plan stated that "the duties, responsibilities, and functions of the Commissioner of Lighthouses shall be vested in the Commandant of the Coast Guard."[26] Ironically, three years previously, in 1936, the Lighthouse Service claimed that it was the "most decentralized of the government services."[27] Robert Erwin Johnson, a historian of the U.S. Coast

When construction crews came to erect the Cape Flattery Light Station on Tatoosh Island, Washington, the Native Americans protested because the site was one of their fishing grounds. Native canoes can be seen on the beach.

Guard in the twentieth century, points out that the statement, probably meant to illustrate that the service was free of bureaucratic inefficiency, "may have had the opposite effect by implying that . . . the lighthouse [district] superintendents enjoyed a considerable degree of autonomy, with resultant lack of uniformity and economy."[28] This misinterpretation probably would not have been made had Putnam still held control of the service because of the respect he gained among decision makers during his tenure. His replacement, Harold D. King, did not have the time to gain the support of Congress and the Washington establishment before Roosevelt's reorganization plan began to move. Congress approved the transfer of the lighthouse service to the Coast Guard and it became official on 1 July 1939. This marked the end of the U.S. Lighthouse Service. Ironically, President Roosevelt declared the week of 7 August 1939 as Lighthouse Week in honor of the beginning of the service.[29]

Thirteen U.S. Coast Guard districts replaced the former lighthouse and previous U.S. Coast Guard divisions and sections. The transfer meant that 4,119 full-time and 1,156 part-time civilian employees would be moving into a military organization consisting of 10,164 officers and civilians. By 7 July, the move from Commerce to U.S. Coast Guard Headquarters was completed. Adm. Russell R. Waesche, Commandant of the U.S. Coast Guard, appointed boards composed of three officers in each of the districts. The boards decided which Lighthouse Service employees could volunteer for entry into the military service and what ranks and rates they should receive. Usually, lighthouse keepers would be made chief or first class petty officers, while tender masters and chief engineers would be made warrant boatswains and machinists. Former district superintendents were moved to commissioned commanders or lieutenant commanders, based upon the length of their service. A selected number who held administrative positions became lieutenants or lieutenants (junior grade). Those accepting military ranks were usually assigned to district staffs with responsibilities for aids to navigation. Only fifty-nine commissioned officers and forty-four warrants chose to enter the U.S. Coast Guard's officer corps. All the newly-made officers were junior to the already serving U.S. Coast Guard officers.

The New Dungeness Light Station was established in 1857 in Washington State's Strait of Juan de Fuca. A Cape Cod dwelling with a tower rising through the center was the basic design of the first sixteen lighthouses built on the West Coast.

Portland Head Light Station is the most-often-photographed light station in the United States. The exterior of the station's tower, although undergoing some repairs, remains essentially the same today as it was when first lit in 1791.

The numbers (only 103 former full-time lighthouse personnel out of 4,119 chose to become military officers) graphically demonstrate most of the former Lighthouse Service personnel chose to remain civilian employees.[30]

The U.S. Coast Guard put the best possible face on the new arrangements. Coast Guard historian Robert Erwin Johnson writes that "Admiral Waesche and his subordinates seem to have made a genuine effort to treat the personnel of the former Lighthouse Service fairly." Yet, there were some minor, and major, areas of friction. One interesting case involved Capt. John W. Leadbetter, who skippered the tender *Cedar* and was said to have "a social status in the Territory of Alaska comparable to that of a commissioned officer and is held in high esteem throughout the Territory." The board recommended a lieutenant commander's commission for Leadbetter, but U.S. Coast Guard Headquarters rejected the advice, probably feeling the exception might cause unrest. Leadbetter became a warrant boatswain, but "he was reputed to be the only one who wore the gold oak leaves denoting command rank on his cap visor with his chief warrant officer's uniform."

Almost ten years after the merger, a retired civilian lighthouse keeper wrote: "how the Commission had the heart to think we civilian personnel would ever blend with the 15 & 16 year old Coast Guard men, 'is a huckleberry away above my persimmon' no good blood ever existed between or with either group. The Coast Guard & Rear Admirals to[o] brassy for we common Sailors, Fishermen Oystermen & what have you."[31]

Comdr. Frank J. Gorman, the U.S. Coast Guard's chief financial officer, reported in 1940 that the merger had netted a saving of some $1 million, almost 10 percent of the Lighthouse Service's annual budget. The centralization of administration and depots accounted for much of the savings. As more Coast Guardsmen took over for the highly paid civilian workers, Gorman expected to see even greater savings. Thus, in the vernacular of today—using the bottom line of saving money for the government—the merger was a success, even though it effectively ended one of this nation's oldest federal maritime services.[32]

Tales of Seven Beacons

BEFORE DETAILING SEVEN LIGHT STATIONS, it would be a good idea to have an overview of the variety of light stations. While most Americans may picture lighthouses as round, the towers of light stations can take many forms: cylindrical, conical, square, octagonal, and triangular. As mentioned in chapter 1, tower heights can vary because of geographical location. That is, if a light tower sits on a high cliff, such as at Point Reyes, California, the tower can be small, while low coastal areas, such as at Cape Hatteras, require tall towers.[1]

The general location of lighthouses provides a method of dividing the light stations of the United States into two very broad categories: onshore and offshore. The construction of the towers within these broad categories provides a method of determining various types of light stations. One type of tower within the onshore category, for example, is a wooden light structure. An example of this type is the tower at Plymouth (Gurnet), Massachusetts. Another type of construction is the masonry tower made of rubble stone, such as that at Beavertail Light Station, Narragansett Bay, Rhode Island. Cut stone (or dressed stone) at Sandy Hook, New Jersey, brick at Great Cumberland Island Light Station, Georgia, and concrete are other masonry construction types. Also within this type are tall masonry towers, such as at Cape Hatteras, North Carolina. Yet another type of masonry tower is the offshore wave-swept tower built on rocks above or slightly below the surface of the sea and subject to the force of the sea. The light station at Tillamook Rock, Oregon, detailed in this chapter, is an example of a wave-swept type. Concrete towers began to replace brick towers at the turn of the twentieth century. Towers of concrete in the United States are usually in the west, as they are best adapted to the chance of earthquake damage. The light station at Makapuu Point, Hawaii, also detailed in this chapter, is of this type.

Yet another type of tower within the onshore category is of metal construction. The tower can be cylindrical, conical, square, octagonal, or triangular as in the shapes of masonry towers, and the metal can be cast iron, wrought iron, or steel. The present tower at Cape Henry Light Station, Virginia, is the tallest cast iron tower in the United States. Some metal light towers had an exterior skeletal support for a narrow cylindrical tower. This type of skeletal structure was built both on- and offshore. An example of an offshore skeletal light station is at Sand Key, Florida, discussed in this chapter. White Fish Point Light Station, Michigan, is an example of a skeletal, tubular tower on land.

The offshore light stations are subdivided into four types. A light station built upon a straight-pile foundation had piles driven into the bed of the salt water. Brandywine Shoal Light Station, built in 1828 in Delaware Bay, is an example of this type of light station. The light station was destroyed by ice in the same year. Another form of pile foundation is the screwpile, such as the Drum Point Light Station (discussion follows). The caisson light station is yet another type of offshore construction and receives its name from its method of construction. Lawrence Potts, an English physician and inventor, in 1845 sank a section of hollow tubing from the surface of the sea to the sea floor. Potts placed a large pump attached to the end extending above the water and as the machine pumped water from the tube (the caisson), it also drew up sand that allowed the tube to sink by gravity deep into the sea bottom. The Sabine Bank Light Station, Texas, is an example of this type of tower.

In 1850 J. Hughes, another engineer, found when large rocks stopped the descent of the caisson, air pumped into the tube allowed his men to descend into the caisson and remove the obstructions, which would then allow the tube to continue its descent. In pneumatic caissons, a metal cylinder pushed down a wooden box-like caisson to the bottom, and then water emptied from the area by pressurized air. An airtight shaft built in the center of the assembly could contain a ladder by which men could climb up and down. It provided access to the caisson, and the workers would muck out the debris. Alpena Light Station, Michigan, is an example of a pneumatic caisson type of light station.

Crib foundations are another type of offshore light station. Ships towed the wooden cribs, assembled on shore, to the selected site and then sunk the cribs with stone. This form is found extensively in the Great Lakes. Cofferdams were sometimes used to construct submarine stone foundations. The temporary wooden cofferdam, brought to the site by ship, again would be sunk after being bolted together and sealed. Finally, the water was pumped out. The foundation of the structure consisted of granite. Cofferdams would also be used to build light stations in shallow water or on small rock outcrops and then cement used for the foundation. Spectacle Reef Light Station, Michigan, discussed in this chapter, is of coffer dam construction. The last type of offshore light station is the Texas tower, which is discussed later in chapter 6.

With the above overview of the various types of light stations, it is time to look at our seven structures. The light stations show a national sweep, unusual construction problems, and a variety of types.

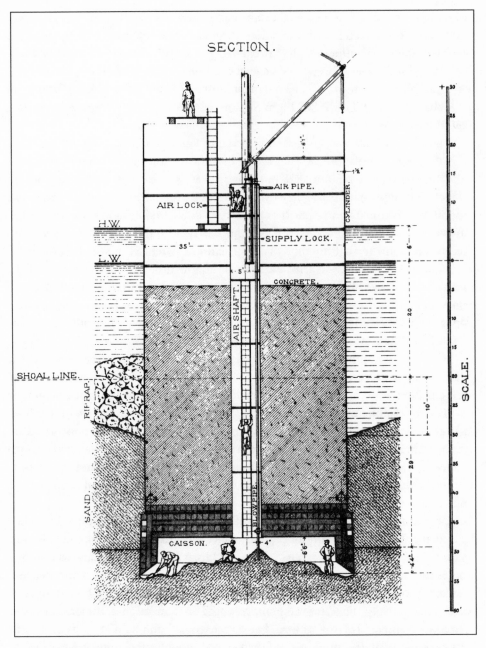

The nineteenth-century technology required for building caisson lighthouses is illustrated in the engineer's drawings for the Fourteen Foot Bank Light Station in Delaware Bay. Initially lit in 1887, this was the first pneumatic submarine-type caisson lighthouse structure in the United States.

CAPE COD (HIGHLAND)

On 6 February 1792, the Massachusetts Humane Society resolved to petition "His Excellency the Governor" on the "necessity of having a Light House erected on some part of Cape Cod. . . ."[2] The Boston Marine Society joined in this request.[3]

The approach to the governor of Massachusetts produced no results, so the marine society took additional steps to establish a light on Cape Cod. In February 1796, the society sent a petition to Congress, which then met in Philadelphia.[4] On 17 May 1796, Congress approved a bill for a light on Cape Cod, Massachusetts, and on 18 June 1796, passed legislation to cede ten acres of land to the United States for each lighthouse site at Cape Cod and Baker's Island.[5]

As indicated in chapter 1, there were no specific instructions regarding how to site lighthouses. Collectors of customs at the various ports also acted as superintendents of lighthouses for their districts, and they usually recommended where to locate the lights. Gen. Benjamin Lincoln held the post of collector for Boston. Commissioner of Revenue Tench Coxe requested Lincoln and Rep. Nathaniel Freeman of Plymouth, who sponsored the legislation, to send advice on where to locate the new light.[6]

Coxe's letter to Lincoln suggested some considerations in selecting the site. Many of the suggestions, interestingly enough, centered more on the land for farming and how the keeper would be employed rather than on the light. This suggests that perhaps the wages of a keeper were so low that farming would be the main source of his income and lighthouse tending a secondary profession. The only suggestion for the actual light was an admonishment that the elevation of the site was "of great consequence" and to employ "an instrument" to find the height.[7]

Lincoln traveled to the area to seek the site for the lighthouse. Isaac Small sold the land for $110. Part of the reason for the purchase centered on the nature of the soil. The land supported farming. The purchase price may also reflect the fact that Small lowered the price in order to be appointed keeper of the light.

The height of land at the site determined that the lighthouse could be a small, squat structure and thus of less cost to a budget-conscious Congress. The light structure, according to Coxe, would be of wood. Lincoln recommended a structure from forty to forty-five feet in height. The actual construction of the structure started with a controversy. General Lincoln announced that he had contracted with Theodore Lincoln, his son, to build the lighthouse.

Commissioner of Revenue Coxe felt General Lincoln's "integrity and candor . . . [would render] satisfactory" any investigation into the arrangement. Perhaps not too surprisingly, Theodore Lincoln did receive the contract. In any case, the proposal submitted by the younger Lincoln called for an octagonal wooden tower forty-five feet to the lantern room, supported on a stone foundation. The contract called for the lantern to be six feet in diameter and eight feet high. The building for the keeper would be one story, twenty-five by twenty-seven feet. A vault for oil, covered by a wooden shed, a well, and a small barn completed the plans for the new station. The final cost for constructing the light at Cape Cod came to $7,257.56.[8]

By the time the Cape Cod light came into operation, there were a number of other lights in New England. At least five lights would mark the way to Boston. This made it necessary to have differing characteristics for the lights so that the mariner could distinguish one light from another. General Lincoln suggested making the Cape Cod station a vertical double light. He then reconsidered and thought the light should have a "blind" that would pass in front of the light once a minute, thus distinguishing it from the fixed lights in the area. The blind, produced by a device known as an elipser, was used in Europe, but not yet available in America. Lincoln contracted with a John Bailey, Jr., who the general considered "the first mechanical genius in this state," to manufacture an elipser. Bailey's contract called for him to deliver the device by 16 October 1797, at a cost not to exceed $500.[9]

Bailey delivered his device on time. Except for a very brief period in the beginning, the elipser worked as promised. Cape Cod "probably has the distinction of housing the first elipsing mechanism in an American lighthouse. . . ." The light officially went into service in 1797 and was the twentieth lighthouse in the country.[10]

The lighthouse developed some problems, such as panes of glass falling out because of the shrinkage of the wood in frames in the lantern room and leaks in the roof. Apparently the repairs were made, as no major alarms about the light came about until 1809, when Keeper Small began to list several problems from broken windows to peeling paint. One of the major complaints concerned the condition of the light itself, and the reason for the poor light centered on the elipser. The device, according to Gen. Henry Dearborn, who replaced General Lincoln as the collector at Boston, reduced the quality of the light. Further reducing the light was the spray from the sea that "intervenes to obstruct the light."[11] One observer defined the problem of the elipser: "As the skreen is continually turning, the light is *full* only for a single moment in the course of each evolution; it is also totally eclipsed but for a single moment; but, during all the time between, it is no more than an obscure and imperfect light, with greater or less difficulty distinguished."[12]

In April 1811, Winslow Lewis announced he now had his Argand lamp ready to install in the Cape Cod light. Keeper Small removed the elipser in preparation for the new lamp. Lewis examined the light structure and found it so defective that it was impossible to install the lamp. Strangely, Lewis noted that the structure "now is much to[o] high and would be seen at a much greater distance if the light was lower. . . ."[13] The lowering of a light to make it be seen farther seems to contradict common sense. The brightness of Lewis' lamps, however, seemed to make it possible to reduce the height of the structure.

Lewis signed an agreement on 17 June 1812 to do the work required to reduce the tower by seventeen feet and to install a lantern ten feet high, for a cost of $2,000. The contract was completed by February 1812.[14]

In the late 1820s, the wooden structure was nearing the end of its useful life. The

Treasury's annual report for 1828 noted that the "whole structure is very imper-fect—is easily wracked by the winds, which shakes the lantern so much as to break out the glass very frequently."[15]

Fifth Auditor Stephen Pleasonton, now in charge of the country's lighthouses, wrote to the superintendent on 27 June 1831 of his decision to rebuild the Cape Cod Light Station. On 22 July 1831, the superintendent signed a contract with Lewis to build the new station. The new light structure would be a round brick tower thirty-five feet high, with a diameter of twenty-two feet at the base. A new keeper's dwelling, also of brick, was to be twenty-six by twenty-eight feet. Lewis would receive $3,993 for the project. A National Park Service study of the light-house says that the exact location of the new structure is questionable, but felt to be close to the original tower. Likewise, the exact date of the completion of the new station is questionable, either 1831 or 1833. The Treasury noted an expenditure of $4,162.78 for the light in 1831.[16]

The conflict between I. W. P. Lewis and his uncle, Winslow Lewis, and Pleasonton can be viewed in microcosm at Cape Cod. The younger Lewis undertook a contract with Pleasonton to refit the Cape Cod light with a Fresnel lens and completed the work in August 1840. Pleasonton began to question the cost of the work, as Lewis's uncle undertook similar work at Cape Henlopen for a much lower cost. Further, the new lamps at Cape Cod used more oil, hence cost more, and—to someone who watched the bottom line as did Pleasonton—this was another point of indictment against I. W. P. Lewis. Pleasonton, late in 1842, ordered the removal of the new lens and the replacement by Winslow Lewis's Argand type. According to one report, there was no difference in the brilliance of the light between the Argand and Fresnel. This, of course, pleased cost-conscious Pleasonton. He could report that his actions saved money for the government.

The case of the Cape Cod station is only one part of the long feud between the Lewises and Pleasonton. I. W. P. Lewis conducted his own investigation of the lights of New England. One of the stations he particularly focused his wrath upon was Cape Cod, the station that his uncle completed some ten years previously. Lewis's list of poor workmanship is long. He noted that the "window frames and staircase" were rotten and could be "pulled out by hand." The removal of the staircase "brought down a portion of the inner wall." These, and many other items, Lewis carefully recorded. One can almost see I. W. P. Lewis smiling as he penned the charges. Pleasonton, however, withstood the attack until replaced by the lighthouse board in 1852.[17]

In 1852, the U.S. Lighthouse Board recommended that the Cape Cod light be elevated, improved, and fitted with a first-order Fresnel lens as it was "an impor-tant seacoast position to mark the approaches to Boston bay. . . ."[18] A year later, the board further recommended moving the light south to higher ground as "the present light is not seen in approaching it from the south over hills, woods &c which are situated a short distance to the south."[19] This move would also enable the board to shut down a triple beacon at Nauset in Eastham. Moving the Cape Cod light to the south of its present location would allow it to be the principal light for the entire eastern side of Cape Cod. Congress approved on 3 August 1854 the "removal of the light-house at Truro (Highlands), Cape Cod, to a proper site, and for fitting the same with the most approved illuminating apparatus, and to serve as substitute for three lights at Nauset Beach."[20]

In the end, however, the board decided to rebuild a higher light on the same property and near the same location as the second light, the additional height and the improved Fresnel lens apparently being enough to cause the board to reconsider moving the station farther south.

Preparations for the new tower continued through 1856, and the actual construction of the tower, an assistant keeper's dwelling, and passages connecting them to each other and the principal keeper's dwelling took place in 1857. The new principal keeper's quarters was built in 1856.[21] The new circular white-brick light tower stood sixty-six feet, two inches. Wall thickness at the base measured forty-two inches and at the parapet thirty inches. The light had a first-order Fresnel lens. It officially went into operation in 1857.[22]

In 1873, a fog signal was established at the station. In 1900, the U.S. Lighthouse Board decided to change the characteristic of the light from fixed to flashing. The new flashing light went into operation on 10 October 1901. The Cape Cod Light Station thereafter flashed white every five seconds, although its characteristics would change from time to time over the years. Many repairs on the station took place at various times, such as the "improvement of the double dwelling occupied by the assistant keepers [that] were completed and a veranda [that] was built on the keeper's dwelling" in 1902.[23] One of the most important changes occurred in 1932, when the light received electricity. This relieved the keepers of their labor-intensive duties, such as trimming the wick and polishing, along with much less climbing of stairs.[24]

In 1961, the U.S. Coast Guard razed the assistant keeper's dwelling and the brick passageways between the buildings. The service then constructed a new duplex for "family quarters." In 1976, the U.S. Coast Guard officially changed the name of the unit to Cape Cod Light Station.[25]

Throughout this period of the third lighthouse rebuilt in 1857, the tower remained in good condition. The end of the manned light, however, came in 1987, when the beacon became automated. In the words of one study on the structure, "it has retained a great deal of its historical integrity."[26] Cape Cod light now sits within the Cape Cod National Seashore. The light is still an active aid to navigation.

The greatest danger to the site was erosion of the cliffs. There was a real danger of the light tower being destroyed: by 1996, the station site sat 110 feet from the eroding 130 foot cliff. The Truro Historical Committee, and other groups, worked for at least ten years to save the light station. On 8 March 1996, the National Park Service announced that the U.S. Coast Guard had selected International Chimney Corporation to move the light tower and attached keeper's quarters. Plans call for the structures to be moved 450 feet inland. This is only the second recent attempt to move a masonry lighthouse. The first successful move was the 1,800-ton Southeast light tower and keeper's quarters on Block Island, New York, in 1994, also completed by the International Chimney Corporation. Work to remove the Cape Cod light began during July of 1996.[27]

CAPE HATTERAS

Far to the south of Cape Cod, another land mass juts out into the Atlantic. Early mariners southward bound from northern U.S. ports to the Caribbean or the Gulf of Mexico kept closer to shore, thereby avoiding the powerful north-flowing Gulf Stream. The area of Cape Hatteras proved to be one of the major hazards in this route. The cape and its shoals reach out into the area ships would take southbound, leaving only a small amount of sea room to maneuver. If a navigator made a mistake in his course to the west, the ship could run aground. A mistake to the east, the vessel would be in the Gulf Stream. This would slow the vessel but, even more important, the weather and sea conditions a ship encountered transiting between the cold waters and warm Gulf Stream could cause danger.

The crumbling cliffs near the Cape Cod Light Station are evident in this photograph. (USCG, courtesy of U.S. Lighthouse Society)

The topography of the Cape Hatteras land area presents yet another hazard to the mariner. Cape Hatteras features low-lying sand enclosing a large area of water, with virtually no landmarks. A sailing ship's captain had to run as close to the shore as possible to gain any type of fix from shore. A mistake could spell disaster. The largest danger to seamen was Diamond Shoals, just to the north of the cape. Clearly, mariners needed aids to navigation in the Cape Hatteras area to provide landmarks—or, more correct, "day marks"—and to help warn ships away from the shoals in the day and a light at night.

In 1794, the U.S. Senate instructed the Treasury Department to look into establishing a light at Ocracoke Inlet. Treasury officials took this opportunity to survey all the North Carolina coast. Tenche Coxe, then commissioner of revenue, reported that most of the maritime interests of the North Carolina coast preferred a light at Cape Hatteras rather than at Ocracoke Inlet. Convinced that a light at the cape would benefit more people, he recommended a light be placed at Cape Hatteras and a small beacon at Ocracoke. Alexander Hamilton, the Secretary of the Treasury, agreed and, in 1794, Congress authorized a beacon for Shell Castle Island at Ocracoke and a lighthouse at Cape Hatteras.[28]

Ten years passed, however, before a light finally shone from Cape Hatteras. This decade long delay centered on finding a builder who would undertake the contract for the lighthouse.

The state of North Carolina first agreed to cede land to the federal government at Cape Hatteras and Beacon Island for lighthouses and then delayed for about six months before ceding jurisdiction over land at Shell Castle Island, instead of Beacon Island. The cession of the land had a three-year limitation, a fact that would cause some complications in the years ahead. Finally, the federal government went about acquiring land for the two aids to navigation. On 29 November 1797, John J. Blount and John Wallace, owners of Shell Castle Island, sold a lot, seventy feet by 140 feet, with the stipulation that the federal government "shall not permit goods to be stored, a Tavern to be kept, spirits to be retailed or Merchandise to be carried on, on said Lott. . . ." One year later, the federal government, on 26 October 1798, purchased four acres at Cape Hatteras for fifty dollars from Christian Jennett, Thomas Farrow, and Joseph Farrow.[29]

Next came the employment of a contractor, and at this stage everything moved at glacial speed. Local people, according to the collector of customs, did not show interest in a contract because they preferred to place their money in farming and commerce "and of course no one has a surplusage to tempt him into [the] more arduous and less profitable enterprise" of a contractor for lighthouses. Coxe printed and distributed announcements in government offices from Maine to North Carolina advertising for bidders to build the lights. The commissioner of revenue received several bids, but he felt many were "dishonest and impudent" and continued his search.[30]

By this time the cession of land by the state had expired, along with the appropriation for the building of the lighthouses. All of this happened just as Coxe selected

John McComb, the builder of the Montauk Point and Cape Henry lighthouses.[31] The expiration of the cession of land caused Coxe to start the process all over. The state legislature renewed the cession to the land at Cape Hatteras in December 1797. Five months earlier, Congress appropriated $44,000 for the two structures. Coxe again started the search for a contractor.

This time Coxe received an inquiry from an unusual source. Henry Dearborn, just completing his second term as congressman from Massachusetts, sent word that he might be interested in contracting to build the two aids to navigation in North Carolina. Not long after an exchange of letters between Coxe and Dearborn, the ex-congressman and two other parties bid on the contract. In September, President Adams received the bids for approval. Coxe was so sure that Dearborn would receive the contract that he ordered the collector of customs in Edenton, North Carolina, to purchase the land for the two lighthouses.[32]

Coxe left the office of commissioner of revenue before the president took action on the contract. William Miller, Jr., took over the commissioner position and looked at the contract in a different light. Miller felt the contract too vague on some charges, such as transportation. This vagueness, the commissioner felt, favored the contractor. Following Miller's recommendation, President Adams disapproved the contract in February 1798. The commissioner, however, wrote to Dearborn offering to help rewrite the contract to weed out the vagueness. Miller also reestablished contact with McComb suggesting that he still might like to make a bid, and he also contacted the collector of customs in Savannah, Georgia, asking if anyone there would be interested. No contractor responded to the query. In the fall of 1798, Dearborn again submitted a bid, this one finally accepted by the president and, on 9 October 1798, the commissioner of revenue forwarded the bid for Dearborn's signature, which Dearborn signed on 31 December 1798.[33]

The site finally chosen for the construction of the Cape Hatteras lighthouse cost fifty dollars and enclosed four acres. Plans called for a light tower of stone and brick, with an octagonal shape rising ninety feet, plus a twelve-foot lantern. The entire station would have an oil vault, shed, and a frame two-story keeper's quarters measuring thirty-four feet by sixteen feet.[34]

Toward the end of August 1799, Dearborn's ship and workers arrived at the Outer Banks area from Boston. After a brief stop at Shell Castle Island, work on the Cape Hatteras site began by building the keeper's quarters. The work progressed satisfactorily until winter, when the crews left the Outer Banks for the season.[35]

The next year, Dearborn gathered another work crew and headed south to work on the foundation of the Cape Hatteras light. By May, the laborers had sunk a hole eight and one-half feet and laid the first course of stone for the structure's base. The first week in July saw the tower twelve feet above the water table and the flooring for the second story in place.

Dearborn's rapid progress, however, slowed. A sickness, described only as a "fever," struck the work force and, before the end of July, thirteen of his workers were struck down and one died. Dearborn fled the area, leaving Dudley Hobart in charge. By fall, Hobart sent the laborers home to regain their health. The collector reported the Cape Hatteras tower situation in the typically difficult-to-understand official language: "The Foundation of the Light House at the cape has been sunk 9 feet below the water Table (or surface of the Earth) is commenced 29 feet in diameter, to the height of 4½ feet from the commencement is solid and from thence to the water Table the Wall is 9 feet thick, and the first story is raised 12 feet from the bottom of the water Table to the top of the Beams. The outside is laid with hammer dressed Stone 10 inches thick and from 1 to 4 feet long, of a firm and durable kind."[36]

In 1869, however, when work on the second Cape Hatteras tower began, a superintendent of construction noted that the first tower's building material consisted of "slate stone varying in thickness from 2 to 6 inches; put up with lime mortar and only one foot larger at bottom of foundation than at base of foot." The diameter of the wall, according to the superintendent, measured twenty-six feet six inches and the wall had a thickness of five feet, six inches. Furthermore, the "lime mortar [was] very soft and stones of foundation laid on the sand with no timber under them."[37]

Work continued on the lighthouse and, by August 1802, Hobart reported work completed on the Cape Hatteras light. For some unknown reason, however, the light did not go into operation until 29 October 1803. The first keeper of the light, Adam Gaskin, was "undoubtedly . . . a good party man."[38]

An experiment in lantern illuminants took place at Cape Hatteras. Someone proposed porpoise oil for the light, probably because of a large supply of the oil in the area. The trial continued until at least 1810 and then stopped, one of the major reasons being the high cost of the oil. Nevertheless, porpoise oil was the choice for emergencies.[39]

Problems soon arose at the new Cape Hatteras light. The oil vault proved too small, requiring a new one to be installed. One of the infamous Outer Banks storms struck on 6 October 1806 and damaged the lantern so badly the keeper could not display the light. In January 1809, fire destroyed the lantern.

In 1810, the keeper's quarters and the tower needed work. A new wire netting around the lantern to keep birds away from it was on the schedule. The two most serious problems were fires in the lantern and erosion. Erosion at Cape Hatteras would prove the most persistent problem, continuing to the present day. The sand hill on which the tower rested blew away in 1810. The first actions taken against erosion at Cape Hatteras are not known, but the light continued in operation. Fires continued to plague the lantern.

Cape Hatteras light came under the care of Winslow Lewis, who received the contract to install his lamps into the light. The light received eighteen lamps and reflectors. Even with the improvement of Lewis's lighting system, complaints about the Cape Hatteras light surfaced. One vitriolic letter came from a John D. Delacy

The striking spiral-band pattern of the Cape Hatteras light tower is now a symbol of the National Seashore wherein the tower sits. The Cape Hatteras structure is the tallest masonry lighthouse tower in the United States.

of Beaufort, who listed innumerable charges against the keeper, Joseph Farrow. Lighthouse historian F. Ross Holland concluded that, "one would like to think that Delacy had the purest of motives" for writing such a stinging letter against Farrow, but "since he transmitted a letter from William Bell, a retired ship's master who was a fellow townsman of Delacy's" and since "Bell was applying for the job of keeper of the Cape Hatteras light," the letter might be "suspicious."[40]

The complaints about the quality of the light continued. After Stephen Pleasonton took over the control of the lights in 1820, a letter published in a local newspaper and signed by several ship masters and merchants complaining about Cape Hatteras Light Station caused him to remove Joseph Farrow. The collector then nominated Pharoah Farrow, who took office in April 1821.

The Cape Hatteras Light Station by now needed repairs, most of a routine nature, except whitewashing the tower. Another serious problem was that sand again was blowing away from the foundation causing erosion. Samuel Treadwell, collector of customs, suggested placing tar around the foundation to prevent the sand from blowing away. The collector, obviously not a man with a nautical background recommended not whitewashing the tower, as it would have to be repeated in time. Treadwell failed to realize that the white tower also acted as a day mark for mariners. To compound the mistake, Treadwell's superiors, first the collector of customs and then Pleasonton, also failed to see this mistake.[41]

By the early 1830s, the lighthouse station began to deteriorate. Pleasonton authorized the building of a new keeper's quarters, as long as the cost did not exceed $1,500. The fifth auditor also authorized more land "at a reasonable price" for the station if the new building warranted it. The new quarters apparently required the land, for not only were "forty acres more or less" added to the station, but new keeper's quarters also went up.[42]

A complaint about the quality of the light that was lodged in the newspapers led to Pleasonton rebuking the keeper through the collector of customs. Pharoah Farrow kept the light, but did not care to live at the station. Farrow solved this problem by hiring several African Americans to tend the light. When news of this reached Pleasonton's office, he removed Pharoah Farrow from the station and appointed Isaac Farrow to tend the light.[43]

Cape Hatteras Light Station continued to deteriorate, both structurally and in the quality of the light. In 1835, Winslow Lewis, under contract, set about refitting the light. The original contract called for eighteen lamps for the light. Lewis's refitting of the Cape Hatteras light seemed to improve the quality of the light until the next major period of change.

Five years after Lewis installed the new lamps, the collector of customs reported that the lamps and reflectors were useless. Pleasonton, of course, immediately lashed out at the keeper, threatening to remove him, because any failure of the lamps had to be his fault, as Lewis "always has done good and substantial work." The collector received orders to investigate the charges and to hire workmen to do any necessary repairs. The keeper, Isaac S. Farrow, however, remained on duty. The workmen made the necessary repairs.[44]

According to the fifth auditor, things went relatively well at the Cape Hatteras Light Station over the next few years. Isaac Farrow died, a new keeper came aboard, Lewis accomplished some needed repairs in 1845, and some work took place on the recurring erosion, but, in general, Cape Hatteras Light Station remained in good order. Then, the administration of the lights passed from Pleasonton to the U.S. Lighthouse Board. The investigation of 1850 that eventually led to the establishment of the U.S. Lighthouse Board did not paint a very good picture of the Cape Hatteras Light Station.

In the late 1850s, G. W. Blunt issued a notice to mariners based upon a friend's investigation of the Cape Hatteras light. Blunt reported a "notoriously" bad light that was "badly kept." Everything about the lighting apparatus showed improper care. The keeper allegedly employed a slave to tend the light, and at one time the paid keeper absented himself from the area for "three months."

The collector of customs, R. H. J. Blount, wrote a defense of the keeper. Holland, however, points out that the defense really amounted to "excuses to get the keeper off the hook." For example, the matter of painting or whitewashing the tower reared its head again. Blunt's report stated that the tower had never seen paint or whitewash. Collector of Customs Blount responded that if whitewashing was really necessary he did not understand "why it has been omitted for so many years." According to Holland, the responses of Blount point out the failure of the fifth auditor's tenure of the lights: "the local superintendents of lights seldom had any appreciation for the role of lighthouses and rather than initiate a little action to keep lighthouses in good condition, they tended to act only when stimulated by criticism."[45]

One of the criticisms of the light was that it was obscured. Pleasonton felt the tower was too high and this caused the problem, as the haze on the coastline of the south obscured lights around a hundred feet in height. The fifth auditor suggested lowering the tower and selected Benjamin Isherwood, the chief engineer in the U.S. Navy, to look at the light station.

Isherwood found what Pleasonton suggested. A mist or low stratus, formed because of the warm Gulf Stream, hovered at about one hundred feet. Lowering the tower would solve the problem, but it would also reduce the range of the light. Because of the meteorological conditions, "the obscurations of Lighthouses due to this cause are therefore irremediable and must be endured as the inevitable result of natural laws beyond the power of man to abolish or modify." This is the type of report that Pleasonton appreciated, for it allowed him to remain with the status quo.[46]

While all this was taking place, an independent inspection of Cape Hatteras Light Station by the nascent U.S. Lighthouse Board took place. The examination first described the light and then went on to detail everything wrong with the light and structure. The report stated that the tower rose ninety feet from base to lantern. The lantern contained fifteen Argand lamps, each fitted with a twenty-one-inch reflector and all reflectors were "much worn." The characteristic of the light was fixed white, visible in clear weather for twenty miles.[47]

The light won no praises for its quality. Lt. David D. Porter, U.S. Navy, had written in 1851 that he had so little confidence in the light he preferred to navigate by the lead. "The first nine trips I made," he related, "I never saw Hatteras light at all, though frequently passing in sight of the breakers; and when I did see it, I could not tell it from a steamer's light, excepting that the steamer's lights are much brighter. . . ." The comments of other mariners proved just as devastating. Lt. H. J. Harstene, U.S. Navy, for example, wrote that the light at Cape Hatteras, "if not improved, had better be dispensed with, as the navigator is apt to run ashore looking for [it]."[48]

The new U.S. Lighthouse Board recognized the importance of the Cape Hatteras Light Station and asked the U.S. Army to help with technical problems at Hatteras and other locations. An engineer made his requested inspection and submitted his report on elevating the tower and installing a first-order Fresnel lens. Congress appropriated the funds, and a brick addition elevated the focal plane of the light to 150 feet above sea level. A Fresnel lens, with a characteristic of flashing white, soon shone out over the Atlantic for at least twenty miles. The board recognized the tower made an excellent day mark and painted a color scheme to make it recognizable: the first seventy feet of the tower were whitewashed and the remainder was painted red. The improved light continued without undue problems until its next major crisis—the Civil War.[49]

Confederate troops moved out onto the Outer Banks on 20 May 1861, shortly after North Carolina seceded from the Union. They seized the light station and extinguished the light. Union forces retook the Outer Banks and wished to display the light as soon as possible. The 3rd Georgia Infantry, however, attacked and drove the Federal troops down the banks. The Union forces managed to regroup at Cape Hatteras. Then, reenforced, the Federals counterattacked and drove the Confederates away.[50]

The U.S. Lighthouse Board, wishing to reestablish the light station, sought assurances that the light would receive protection in case Rebel forces again menaced the area. A letter from the secretary of the treasury to the secretary of the navy addressed the concern and requested protection "from future injury or attack." The navy issued orders to the commander of the blockading squadron to provide coverage from seaward. Similarly, the army replied: "The War Department will cordially cooperate with you in giving the proper instructions to the officers under its command in that region to do everything in their power to protect and secure the light from future injury or attack."[51]

When the U.S. Lighthouse Board decided to relight Cape Hatteras light, they found the Confederates had removed the lens from the tower. When shown again in June 1862, the light was equipped with a second-class lens, which in 1863 was upgraded to a first-class lens.

After the war, the U.S. Lighthouse Board deemed that the Cape Hatteras tower needed a new stairway for the tower. The district engineer, W. J. Newman, however, thought "that the condition of the tower is such that it is not worth the contemplated outlay, but rather that an appropriation be applied for, to build a new tower." Newman went on to list several problems with the tower and underlined the need for construction by pointing out that the "structure is quite out of date and liable sooner or later to a disaster." Apparently, the board recognized Newman's thesis for, in March 1867, Congress appropriated $75,000 for rebuilding the Cape Hatteras lighthouse.[52]

By 1868, the board entered into a contract with Nicholas M. Smith of Baltimore to supply 1,000,000 "'prime dark red' brick" and entered into contracts with other companies to supply all the needed material for the light tower. The tower's height, originally set at 150 feet, now would rise to 180 feet because of "the interests of commerce." Work on the tower continued at a steady pace on the new site, some six hundred feet northeast of the old tower. Some problems naturally cropped up. One concerned workers' health. Fever struck down the laborers in early 1869. Surgeon Col. James Simons recommended "a daily issue of 1½ ounces of whiskey and 5 grains of quinine to prevent intermittent and remittent fevers."[53]

On 17 September 1870, the keeper of the new Cape Hatteras Light Station displayed the first-order light on the tallest masonry light tower in the United States. The light measured 190 feet above mean low water. The brick tower, covered with "cement wash to protect it" against the weather, had a color scheme of red on the upper parts, "the lower part (projected against the foliage in the rear) colored white. . . ." A charge of dynamite destroyed the old tower.[54]

The workmanship on the new tower proved excellent, as the history of the light well into the twentieth century shows only routine maintenance took place. In 1873, to make the tower a better daymark, the structure received its distinctive black and white spiral bands.[55]

The next major change to take place in Cape Hatteras' light occurred on 1 July 1912, when an incandescent oil–vapor lamp came into operation. The use of this lamp increased the intensity of the light from 27,000 to 80,000 candlepower. No further change took place in the illumination of the light until 1934 when electricity became available. The installation of two Kohler generators and a bank of batteries in the building formerly used as an oil house allowed an electric bulb to replace the lamp.[56]

From 1919 to the present, there has been one major concern about the Cape Hatteras lighthouse: the relentless encroachment of the sea. In 1919, the surf washed upon the shore no less than a hundred yards from the light, as opposed to a mile when the light went into operation. This caused the beginning of many types of efforts to stabilize the dunes in the area. For example, shrubs and grasses were planted and dunes constructed in the hope it would slow the sea. The area did stabilize, but "probably more as a result of natural fluctuations than these artificial measures." After things seemed to return to normal in 1920, life continued in the traditional manner at the lighthouse. Over the years, however, the sea continued its inexorable advance toward the tower despite the efforts of many concerned individuals and organizations.[57]

By 1936, the Atlantic lapped at the base of the tower and the service had to make a decision about the lighthouse. On 15 May 1936, Keeper Unaka B. Jennette supervised the extinguishing of the Cape Hatteras light. In 1939, the U.S. Coast Guard took control of the lights. A metal framework went up a mile and a half away from the abandoned tower in Buxton Woods. This would seem to be the end of the story of the Cape Hatteras Light Station, but events would prove otherwise.[58]

In 1946, a shipwreck nine miles from the old light tower helped restore the original Cape Hatteras light tower. The yacht *Nautilus* used the old tower for a daymark,

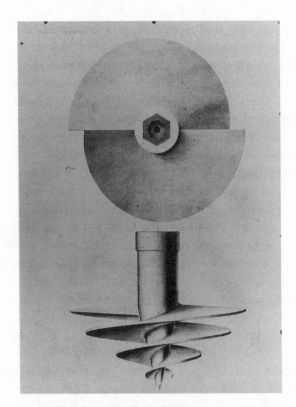

The working end of a screwpile. The screwpile lighthouse proved ideal for holding in muddy, sandy bottoms in certain East Coast areas and among the coral reefs off the coasts of Florida.

unaware of the new light almost two miles away. In the dark the skipper used the new light for navigation, thinking it was on the old tower. As a result, the *Nautilus* slammed into the beach and was lost in the surf. This caused the U.S. Coast Guard to reevaluate the aid to navigation, especially since another dune stabilization program left the tower at least a thousand yards from the sea. In December 1948, the National Park Service agreed to retain title to the lighthouse, but allowed the U.S. Coast Guard a use-lease for twenty years with an option to renew at the end of that time. Restoration began on the ill-treated tower. A forty-inch light beacon with a fourteen-hundred-watt bulb replaced the badly damaged Fresnel lens. The automated restored light went back into operation on 23 January 1950 and produced 250,000 candlepower, visible for twenty miles in clear weather.[59]

The last act of the Cape Hatteras Light Station story is uncertain. That uncertainty lies in the approaching sea. By 1975, the Atlantic washed within about 775 feet from the tower. Five years later, the sea washed away the final remains of the old original tower. By the autumn of 1980, a short seventy-foot distance separated the tower from the sea. Eventually, what to do about a tower that now symbolized a National Seashore came down to two hotly debated options. One consisted of building a large revetment around the light. The revetment would be twenty-three feet above sea level, extend another sixteen feet below the surface of the Atlantic, and its walls would be of one-foot-thickness and reinforced by a twelve-foot-thick riprap of boulders and five-ton concrete pods "reaching out like a talus slope fifty feet from the base." In short, this arrangement could form an island around the light. A walkway would allow visitors access to the structure.[60]

In December 1989, however, the National Park Service opted for the plan to move the Cape Hatteras light tower. Basically, the course of action called for hydraulic jacks to lift the tower, insert needle beams under the tower, and then move the entire structure some five hundred feet. Once started, the operation would take four months. As of July 1996, the National Park Service had not moved the Cape Hatteras tower. Some see this lack of action as an implication that the National Park Service "really intends to let [the tower] fall into the sea." Others point out that the National Park Service is pursuing the best possible course of action: fighting the sea until the "last resort," as this prevents any possible damage to the tower from the attempt to transport the structure. The story of Cape Hatteras light tower remains on hold as humans try to battle the forces of the sea.[61]

DRUM POINT

To the north of Cape Hatteras, in the majestic Chesapeake Bay, a unique type of lighthouse became a common sight. Lt. A. M. Pennock, of the U.S. Lighthouse Board, after surveying the Chesapeake Bay, noted that "a small light should be placed on Drum Point, inside of the Patuxent River." In heavy weather, vessels of all types took shelter in the lee offered by the point, and in doing this several craft had "brought up on the spit." Congress authorized $5,000 on 3 August 1854 for "a light-house on Drum Point, entrance of Patuxent River." The appropriation languished for some eighteen years. On 7 August 1882, Congress again authorized money for aids to navigation in this part of Chesapeake Bay. This time, however, the authorization called for the "establishment of two range lights at the mouth of the Patuxent River." This plan also miscarried. The U.S. Lighthouse Board noted the "smallness of the appropriation," which caused the board to go ahead with a light-house as originally planned.[62]

The U.S. Lighthouse Board decided the light should be of the screwpile type. Drum Point Light Station would resemble a large six sided cottage on tall metal stilts. Screwpile lights were especially useful where a light would have to rest on shoals or coasts with sandy, muddy bottoms. The Chesapeake Bay area proved ideal for this type of structure. Of the seventy-four lighthouses built on the bay, screwpile lights totaled forty-one. The history of the development of the screwpile form of lighthouse is interesting.

In 1773, John Phillips, of Liverpool, England, decided to build a lighthouse on

Smalls Rock in the Bristol Channel off the west coast of Wales. Interestingly, a musical instrument maker by the name of Henry Whiteside designed the lighthouse. The light would sit on piles, allowing the sea to pass through rather than slam into a masonry tower. The wood light sat on nine oak piles anchored by metal rods driven into the rock. The light stood for some eighty-five years.[63]

Alexander Mitchell, the son of the inspector general of barracks in Ireland, became an engineer who, at the age of forty, was blinded. He continued to operate his Belfast company with the help of his son. Mitchell patented a cast-iron screwpile in the 1830s. The end of the metal screwpile resembles a metal auger and is designed to bite into the bottom and gain "a much greater holding power" than any other type of pile "then in use." Mitchell went on to design several lighthouses in England using his screwpile.

As the building of lighthouses in the United States moved away from New England and its traditional masonry or wooden frame towers, certain locations

Boaters must have done a double take at the sight of barge, crane, and lighthouse sailing past. (Calvert Marine Museum, Solomons, Maryland)

seemed fitting for the screwpile design. The first light of this type in the United States, designed by Maj. Hartman Bache, of the U.S. Army, was placed at Brandywine Shoal, in Delaware Bay, in 1850. The Brandywine Shoal has two "firsts": pile structure and screwpile. The structure replaced a lightship. Screwpile lighthouses proved excellent for use in protected water along the Eastern Seaboard and into the Gulf of Mexico region. The screwpiles proved ideal for holding in the coral reefs in the Florida area. It was the Chesapeake Bay area, however, that held the most lights of this type. In 1894, the U.S. Lighthouse Board became concerned about the real danger to lights of this type: ice. The dangers of ice caused the board to switch to caisson-type structures wherever the danger of ice existed.

Work on the light station at Drum Point finally began on 17 July 1883. The ten-inch-diameter wrought-iron piles, manufactured by Allentown Rolling Mills, Philadelphia, had three foot-wide auger flanges. Each of the seven piles had to be bored into the Chesapeake's bottom by hand. Once the piles were set, then the precut foundation came next. The workers assembled the parts by matching the numbers on each piece. This accomplished, now work could begin on the house, or "cottage."[64]

The house—built of wood, six-sided, and one-and-one-half stories—had mortised and tenoned joints and was sheathed with weatherboard. One gallery surrounded the outside of the main floor and a second around the outside of the lantern room. Two iron ladders led to the water's edge from the main gallery. The cottage's color scheme consisted of white paint on most of the wood, with red on the metal roof and pilings. Below the cottage, a platform between the pilings held oil, coal, and wood. The light station stood in ten feet of water. The shallow depth of the water and the prefabrication are two reasons for the completion of the Drum Point Light Station in just thirty-three days and for a cost of $5,000.

Drum Point Light Station received a fourth-order Fresnel lens. The purchase price of the lens, $1,200, made up one-fifth of the cost of the station. The focal plane of the light was forty-seven feet above the bay and visible thirteen nautical miles away from a ship's deck fifteen feet above the water line. Drum Point Light Station produced a fixed red light by having a red chimney over the lamp. The light covered only a 270 degree arc, as the remaining 90 degree arc faced land. At some time before 1909, a white chimney replaced the red, and three sheets of ruby-red glass were attached to the inside of the lantern room windows. This created three red sectors, with white in between. By staying within the fixed white sector, a mariner could navigate into the Patuxent River from the bay.

Drum Point Light Station also had a fourteen-hundred-pound fog bell, manufactured by McShane Bell Foundry, Baltimore, in 1880. The fog bell pealed twice at intervals of fifteen seconds during thick weather. The bell was struck using a clockwork mechanism. One keeper reported that "foxes came down to the beach to bark at the fog bell on foggy nights." Benjamin Gray, the first keeper of Drum Point Light Station, officially first displayed the light on 20 August 1883.

Living conditions were not ideal, but they would be considered adequate for the time. To obtain fresh water, rain water was collected from the roof and piped into four two-hundred-gallon cisterns located with the cottage. During dry periods, the keeper walked a mile to a house, obtained water by the bucket, and carried it back to the light.

Even the keepers of screwpile lights located in protected waters faced dangers from the elements. Ice presented a hazard, but sometimes even storms could make life chancy. On 23 August 1933, Keeper John J. Daley reported that a severe storm with seas "at least 15 feet high" struck the station and flooded all the rooms on the main floor. The waves carried away the station's boat from the main floor. A large tree carried by the waves lodged against the pilings, causing other driftwood to collect against them. After the storm, Keeper Daley wrote that the only way he could get ashore was "to swim."

Compared to the Cape Hatteras Light Station, Drum Point was located in an almost urban environment. Solomons, Maryland, sat only two miles away by water from the station. Keepers could come ashore easily and some kept a garden on the nearby point of land.

Life, in general, passed uneventfully at the Drum Point Light Station. The major change at the light station, before the Bureau of Lighthouses became a part of the U.S. Coast Guard, came in 1932. Then, the incandescent oil–vapor lamp became a part of the station, increasing the candlepower and the range of the light.

The Drum Point Light Station received electrical power on 5 August 1944, and, on 6 February 1950, much of the labor-intensive work of the keeper's job changed when one-hundred-watt light bulbs replaced the incandescent oil–vapor apparatus. Then, on 1 March 1960, the light became automated. By 1960, nature played an important part in the history of the light station. Sand began to build up around Drum Point so that by the early 1960s the light no longer sat in ten feet of water and 120 yards offshore. Now the structure stood high and dry at low water. On 6 September 1962, the light was decommissioned and replaced by a new pile structure.

The former Drum Point Light Station now stood exposed to the tender mercies of vandals. Photographs taken after the decommissioning show a building with peeling paint and the effects of vandals. The state of Maryland at first planned to restore the light and open it to the public. Public access to the former screwpile lighthouse, however, was blocked by miles of private property. The state returned the site to the General Services Administration (GSA). The site continued to deteriorate. John Hanson, a former keeper of the light, said that he avoided the Drum Point beach because he "could not bear to see the condition of the structure."

In 1966, the Calvert County Historical Society undertook the acquisition and restoration of the old screwpile lighthouse. The society began the long and complicated voyage through the maze of local, state, and federal bureaucracies to complete their goal. Seven years later, the goal was still beyond the society's grasp. In an attempt to at least protect the structure, the society began the work needed to place the lighthouse on the National Register of Historic Places.

The former Drum Point Light Station now rests on the grounds of the Calvert Marine Museum for all to enjoy. (Calvert Marine Museum, Solomons, Maryland)

Then, in 1974, the Calvert County Historical Society learned that GSA would let the society have the lighthouse, but not the land it sat upon. With help from the Calvert County government, the lighthouse now belonged to the historical society and apparently none too soon. As G. Walther Ewalt, then the president of the society, said, "Vandals [had] already set the lighthouse on fire, attempted to steal the large bell, stolen the brass lens stand and have ripped all doors from their hinges—even knocked out the railing. It [looked] terrible."

The Calvert County Historical Society obtained a grant of $25,000 from the state of Maryland to move the Drum Point Light Station. In March 1975, a large barge with a steam-operated crane and a 110-foot boom, towed by a tug, made its way to Drum Point. To move the barge close to the light, the tug had to use its screws to force water to cut a channel near the light. After a great deal of effort, the legs of the screwpile light were cut and the lighthouse lifted from its foundation. Residents and boaters in the area had the unusual experience of seeing a strange procession. First came two tugs pulling a large barge with a steam crane; gently swaying from the crane's boom was a lighthouse. A large crowd greeted the procession after the half-hour move and watched as the crane lowered the lighthouse onto the grounds of the Calvert Marine Museum.

A donation helped the museum stabilize the exterior of the lighthouse. Other donations and funds helped with the costs in restoring the structure. The dedication of the restored screwpile lighthouse came on 24 June 1978. The work of everyone involved in the project helped to prevent the loss of one of the historic structures of the water world of Chesapeake Bay. Just as important, the preservation of the old screwpile lighthouse at a site that allows the public access to the light will ensure that a portion of maritime history will not fade away.

SAND KEY

Stretching in a gentle curve southwestward from the southern tip of Florida lie the Keys, an area of shoals, reefs, and low islands that are difficult for the navigator to see until the ship is almost upon them. The light station at Sand Key, located seven and three-quarters miles southwest from Key West, is a good example of the problems involved in building lighthouses in the region.

The first light station on Sand Key consisted of a stone tower with a keeper's house nearby. When the light first went into operation in 1827, the key was a sandy island. The Florida Keys lie directly in the path of hurricanes and, over the years, several large storms played havoc with the light stations on Sand Key. In 1846, a severe hurricane struck the key. Rebecca Flaherty, who had taken over the keeper's duties from her husband, John, in 1830, took her five children into the tower for refuge, but the structure was swept away—killing everyone in the light tower.[65]

The Treasury Department purchased a lightship to mark the location until a new lighthouse was put into operation. I. W. P. Lewis, the controversial nephew of the even more controversial Winslow Lewis, designed a screwpile lighthouse and supervised the placement of the piles. Lt. George G. Meade, then of the Topographical Engineers, later to lead the Union forces at the Battle of Gettysburg, took over the work in May 1852, when Lewis was "obliged to suspend his operations for want of funds, and who was relieved from the charge of the work owing to his appointment being in conflict with an Act of Congress." Lieutenant Meade awaited funds that did not become available until December 1852 and then he placed William Dennison, a civil engineer, in charge of the work. Work began on 22 January and was completed by the beginning of August 1853. Dennison's health "failed him" by the middle of July, and he was relieved of duty by J. W. James, a clerk. Lieutenant Meade closed his report on the new Sand Key Light Station by noting that it "was a source of satisfaction that the work was closed and all parties returned to their homes without the slightest accident, or without a serious case of illness." The light station went into operation in 1853.[66]

Lieutenant Meade left four "precautions" for future keepers and inspectors in his report. Meade insisted that no structure of any kind be permitted under the dwelling "for the free passage of the wave should the Key be submerged." The foundation piles needed to be watched "carefully" and the "whole structure, but particularly the foundation, should be kept well painted, so as to expose it as little as possible to the salt air and water. The braces needed to be checked constantly and the actions of the sea and wind should be reported by the keepers and, if the Key should be completely submerged, how the sand reformed."

Twelve years later, another strong hurricane swept through Sand Key. This time the storm took all the sand from the island, leaving the light undamaged but standing alone in the water.[67]

A lighthouse inspector in 1913 described the light station as a "square pyramidal skeleton iron tower inclosing [a] dwelling and [cylindrical] stairs." The height of the focal plane was given as 109 feet and the tower stood on "16 iron screw piles 14 ft. in coral rock." The first-order lens made a revolution every eight minutes. In order to turn the lens, the clock mechanism for the lens needed rewinding every three and one-half hours. There were four red sectors produced by placing red "panes attached to [the] inside of lantern."

The quarters for the head keeper and two assistants consisted of eight rooms, located at the bottom of the structure, made of "iron circled with wood," and painted brown. Keepers reached the lantern room by a circular stairway from their quarters. Rain water for drinking was caught in four cisterns, each holding twelve hundred gallons.

Sand Key Light Station's inventory included a twelve-foot dinghy that could be fitted with sails. On the west side of the station, a fifteen-by-thirty-foot boathouse on iron piles was connected to the light by a wooden plank walkway.[68]

One description of the Sand Key Light Station rightfully declared the station had "withstood the elements for more than a century proving its stability is a tribute to

The original light tower at Sand Key, Florida, was built of stone in 1827. A storm carried away the tower three years later, though, killing the keeper and her five children. The new screwpile light went into operation in 1853 and is shown as it looked before a fire gutted the lighthouse in 1989. The light was automated in 1938.

the engineer who planned the structure." On 12 November 1989, however, the station suffered an apparent chemical fire that gutted the structure. Cullen Chambers, then director of the St. Augustine Lighthouse, said, "I just don't know how to put it into words, just the sense of loss. It's a tragedy that should be felt nationwide."[69]

A temporary beacon was placed on the lower level of the charred structure. Plans called for a new structure to the south of the former light. This location was objectionable to Reef Relief, an environmental group based in Key West. The group felt "no living coral [should be] destroyed by placement of additional structures on the surrounding coral reef." A temporary site was located in fifteen feet of water "approximately 200 yards away bearing 310 T[rue] from the original light."[70]

Work on the new tower began in 1990. According to Capt. B. W. Hadler, Chief of Aids to Navigation and Waterways Management Branch of the U.S. Coast Guard's Seventh District at Miami, "Local preservation groups urged the Coast Guard to restore the light structure to its original condition. In 1991 funding was made available and plans to rehabilitate the historic structure were effected. Restoration of the structure will be completed and a lantern relighting is anticipated in late July or early August of [1996]. A new state of the art optic, powered by a 12-volt solar system, will be installed. The lights flash characteristic . . . will remain the same."[71]

SPECTACLE REEF

The U.S. Lighthouse Board overcame many engineering difficulties in building lights. One would expect most of these difficulties to be along rocky, isolated coastlines, such as in Oregon, Washington, and parts of Alaska. One of the "notable engineering" feats in the lighthouse service, however, took place in northern Lake Huron at Spectacle Reef. The submerged limestone reef is at the eastern end to the busy approaches of the Straits of Mackinaw, the body of water connecting Lakes Michigan and Huron.[72]

In 1868, the U.S. Lighthouse Board began petitioning Congress for a light at Spectacle Reef, noting that the reef was "probably more dreaded by navigators than any other danger now unmarked throughout the entire chain of lakes. . . ." The U.S. Lighthouse Board ordered that a buoy mark the location in 1868 but noted that the aid was only partially useful as it was only visible "in the daytime."[73]

Many problems would be involved in building a light at the reef. One was that the site sat at one end of Lake Huron and this would expose the structure to the power of waves generated along 170 miles of lake. Another mitigating factor was ice. The structure had to withstand winters on the Great Lakes. Estimates said the site could expect exposure to an ice field thousands of acres in size and often at least two feet thick. When in movement, the ice is an "almost irresistible force." Always conscious of costs, the board also listed as a problem the estimated $300,000 needed for the project. But, the board reasoned, "the wreck upon . . . [Spectacle Reef in 1867] of two vessels at one time involved a loss greater than required to mark the danger. . . ." The board felt that growing commerce on the lakes justified a light.

After a survey of the reef in 1869, Congress authorized $100,000 to begin work and a like amount in the next year to continue the work. The service established a depot, in 1870, at Scammon's Harbor, Michigan, some seventeen and one-half miles from the reef. Survey work revealed that the location proposed for the light station was also the final resting place of the schooner *Nightingale*, along with her cargo of iron ore. This was the only suitable location, so part of the wreck had to be cleared for the light. In charge of the construction was Maj. O. M. Poe, who was Gen. William T. Sherman's chief engineer on the march to the sea in the Civil War.

The plan was to construct a crib at Scammon's Harbor and then tow it to the reef area. The ninety-two-foot crib dam had a central opening of forty-eight square feet. The crib would be sunk by using a ballast of rock. After preparing the site at the reef, on 18 July 1871, the tugs *Champion* and *Magnet* towed a barge with the crib dam to the site, along with a flotilla of other vessels containing supplies and a work force of 140 men. At six-thirty the next morning, the crib was in position and ready for ballasting. "All hands went to work throwing ballast-stone into the compartments, and by 4 P.M. succeeded in getting into it about 200 cords (1,200 tons.)" Rising winds caused a temporary halt to the work, but the next morning the work was completed. By 20 September, the pier built upon the crib rose twelve feet above Lake Huron. Workers' quarters were built upon the pier and the lighthouse schooner *Belle*, which had been serving as temporary bunking facilities, was no longer needed.

A diver began leveling the bedrock within the pier. A coffer dam was then readied and lowered. The coffer dam was a hollow cylinder, forty-one feet in diameter, composed of wooden staves, each four-by-six inches, and fifteen feet long. Internally braced and trussed, hoops of iron circled the outside of the cylinder. The coffer dam was driven down, and the bottom sealed with Portland cement. By 14 October, the pumps had removed all the water from the casement. Stone cutters made their way to the bottom of the coffer dam to the reef and started the work of leveling off the bed rock. To lay the first course of stones for the structure, the high sides of the reef were cut down at least two feet, which involved "a large amount of hard labor. . . ." The first course of stone blocks was set into place by 27 October. According to the official report, the weather had "become very boisterous, with frequent snow-squalls. . . ." The work was shut down and the workers returned to the mainland, except for two men who remained to tend the fourth-order light, established on top of the worker's quarters, and a fog signal consisting of a whistle attached to one of the steam boilers. The two keepers finished their tour at the end of the navigation season and were transported back by the tender *Haze*.

Next year's work began later than planned, as the ice hung on stubbornly. Workers found thick ice on the pier in a "compact mass" in the coffer dam. Once

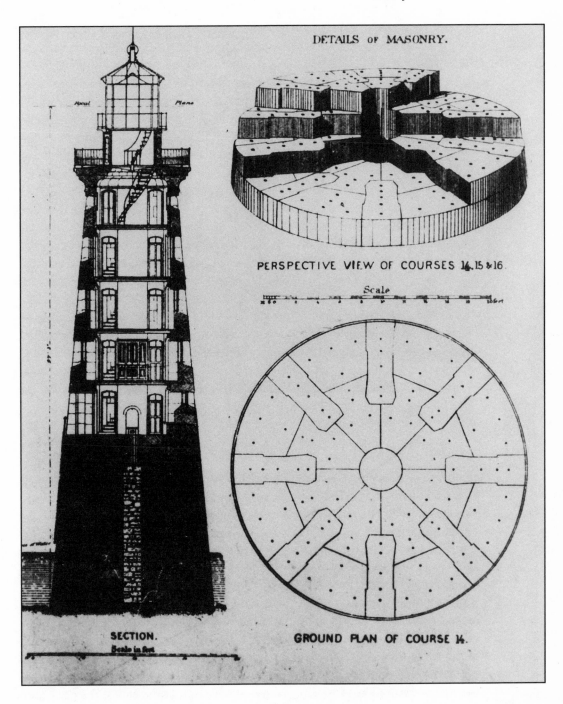

DETAILS OF MASONRY.

PERSPECTIVE VIEW OF COURSES 14, 15 & 16.

Scale

SECTION.

GROUND PLAN OF COURSE 14.

In this drawing of the Spectacle Reef Light Station, Michigan, note how the courses of rock are bolted to the ones below. Each bolt was set into Portland cement that is now probably as hard as the stone.

Spectacle Reef Light Station is described as an "engineering feat" of the U.S. Lighthouse Service.

removed, work began in earnest. The plan was to place a course of granite blocks as a base, then bolt the stone into the foundation with three-foot bolts and twenty-one inches of that length would be into the bedrock, then another course of stone bolted to each other and into the previous course, and so on. Each bolt was set in pure Portland cement. Originally, the granite blocks were to be purchased from a quarry in Duluth, Minnesota, but the contractor "utterly failed" to provide the blocks. Marblehead, Ohio, eventually provided limestone blocks and, by June 1872, six courses were completed.

Spectacle Reef Light Station first displayed its light in June 1874. The light is considered one of the "best specimens of monolithic stone masonry in the United States." The tower rests upon a solid mass of stone that is thirty feet in height. The hollow tower then rises another five stories. Five rooms are within the hollow tower, one above the other, and each fourteen feet in diameter. The walls at the base of the working tower are five feet, six inches tapering to sixteen inches at the cornice. The focal plane of the light is eighty-six feet, three inches from the water.[74]

The displaying of the first light proved difficult. When the first keepers came to the light station in May 1874, they found ice piled thirty feet high, which was seven feet above the entrance. The keepers were kept busy cutting away the ice before they could display the light.[75]

Spectacle Reef Light Station was automated in 1972 and remains an active aid to navigation. The original Fresnel lens was removed in 1982 and is at the Great Lakes Historical Society Museum, Vermilion, Ohio.[76]

TILLAMOOK ROCK

The U.S. Lighthouse Board could rightfully complain that the fifth auditor left them a mess to clean up. The board could also rightfully point with pride to the West Coast of the United States as an example of their own successes. All the construction of the major lights on this coastline was under the supervision of the board until it became the Bureau of Lighthouses. In 1846, "there was not a single lighthouse nor other aid to navigation from the southern end of California to the northern tip of Washington. . . ."[77] In the history of the U.S. Lighthouse Board, only fifty-five lighthouses dotted the coastline of the three Western states. These simple facts do not suggest the difficulties that some lights presented the board. One of the most demanding lighthouses of the West, both to build and for duty required, lies some twenty miles to the southwest of the Columbia River, along Oregon's rugged coastline.

The Columbia River is important to maritime commerce. Unfortunately for mariners, the northern and southern approaches to the river are fraught with the dangerous jutting, rocky headlands of the Washington and Oregon coastline. This was doubly so in the nineteenth century, when seafarers tended to navigate by coasting. That is, skippers hugged the coastline, picking out their navigational fixes from prominent landmarks. The danger in this type of piloting is that sudden shifts of wind or miscalculations can cause a skipper to have his ship driven onto the beach. While today's tourists flock to enjoy the sea stacks and steep, rocky headlands of Oregon's seashore, the nineteenth-century coasting sea captain faced the chance for a very unpleasant end during stormy approaches to the Columbia River.

On 20 June 1878, in response to the loss of life from shipwrecks on the southern approach to the Columbia River, Congress appropriated $50,000 for the construction of a first-order light at Tillamook Head, Oregon. The initial planning recognized that $50,000 would "not complete the work. . . ." An accurate estimate proved impossible at the time, so the board asked for an additional estimated $50,000. The U.S. Lighthouse Board could not realize the portent of things to come when, in 1879, they reported that the construction of the light at Tillamook Head "will be attended with many difficulties. . . ."[78]

The first major hurtle proved to be the location of the light. The original plan called for the structure to sit atop the thousand-foot high Tillamook Head, but the cost to maintain a road to the light was prohibitive. Furthermore, low clouds and heavy fog would make a light perched high on a cliff almost useless. A lighthouse located at sea level was the suggestion of Lighthouse District engineer Maj. G. I.

Gillespie. Gillespie also suggested Tillamook Rock. This uninhabited crag, officially described as an "isolated basaltic rock divided, above low water, into two unequal parts, by a wide fissure with vertical sides running east and west, stands 100 feet above the sea, and has a crest which can be so far reduced as to accommodate a structure not greater than 50 feet square. A comparatively quiet landing can be made on the east side when the sea is smooth. The water on all sides is deep." The report went on to note that the "execution" of the building of a light station at Tillamook Rock would be "a task of labor and difficulty," and would cost a great deal of money. The writer proved to be prophetic.

Gillespie's suggestion at first met with disfavor. The continued shipwrecks on the approaches to the Columbia, coupled with the logic of not building on the headland, however, gave the U.S. Lighthouse Board no alternative but to proceed with the plan to establish the southern light on Tillamook Rock.

In charge of the project was H. S. Wheeler, district superintendent of construction. By June 1879, the weather cooperated, and Wheeler sailed in the U.S. Revenue Cutter Service cutter *Corwin* to the area of the rock. The sea was relatively calm, and the superintendent was placed in the cutter's pulling boat and rowed to near the massive rock. Wheeler has the dubious honor to be the first to learn that a "comparatively quiet landing . . . when the sea is calm" sounded nice in an official report, but at this station the catch would be to find tranquil water. As the superintendent's boat approached the rock, he noticed the seas were calm in all directions, except around Tillamook Rock. Breakers lashed the crag, with whirlpools and eddies in the immediate area. The boat crew and Wheeler decided it was not safe to land and, after making a visual survey from the boat, the cuttermen returned the superintendent to the *Corwin*.

Wheeler found himself ordered to set up a watch at Astoria, Oregon, for the first calm day at the rock and not to return to headquarters until he managed to finish his mission. After long weeks of inactivity, a calm day appeared, and the superintendent was again in a small boat making his way to the rock. Again, only at Tillamook Rock was the sea angry. Wheeler decided to risk a landing rather than return to Astoria and spend more time waiting. The only side of the crag without high perpendicular rock walls was along the eastern approach. Wheeler ordered two cuttermen to attach lifelines to themselves and get from the boat to the rock anyway they could. The boat pulled closer and closer. The craft pitched upward in ten-foot swells and then dropped with express-elevator speed. One can imagine the thoughts of the two "volunteer" sailors as they crouched in the bow of the pulling boat. When the sailors hesitated to make the leap, Wheeler ordered them to get aboard the rock. The men then made the stomach-wrenching hurdle and somehow managed to make it. Now the next obstacle was landing the surveying instruments. The seas, however, began to rise and, fearing a crushed boat, the coxswain backed away from the rock. The two sailors on the rock so feared a stranding that they flung themselves into the cold waters of the sea, and their shipmates pulled them into the boat by their lifelines. Wheeler again retreated to Astoria.[79]

Four days later, the superintendent was again in a boat being pulled out to the eastern approach to the rock. Again, the boat pitched as it neared Tillamook Rock, but this time Wheeler crouched in the bow, waiting for the right moment to hurl himself onto the rock. He was successful. The superintendent then tried to rig lines from the boat to the rock to land his instruments, but failed as the seas, almost as if on cue, began to rise. Wheeler was left with only a hand tape, but set about measuring the rock. With seas building, the small boat was maneuvered close to the rock, and Wheeler leaped into the craft safely, albeit bruised.

Superintendent Wheeler felt that someone with construction experience should make another survey of the rock before construction began. He hired John R. Trewaves, a master mason with experience in building lighthouses off England, to make the survey. On 18 September 1879, Trewaves leaped to the rock, lost his footing, and went into the sea. His body was never recovered. The accident shocked the local seaside community enough that some began to doubt the wisdom of a light on the rock. Nevertheless, Wheeler "immediately" hired A. Ballantyne, a foreman, and eight quarrymen "in order to prevent further delays and to forestall the evil tendencies of public discussions of my plans. . . ."[80]

In the fall of 1879, Ballantyne and his eight men were ready to do battle with Tillamook Rock. As would be the case throughout the history of the light, the party waited for the weather. Twenty-six days later, the group found itself aboard the cutter *Corwin* and near the rock. The construction team clambered aboard a small boat, and the battle to gain a foothold on the rock began. Six hours later, only four men had managed to reach their goal, while the boat had its gunwale stove in and had sprung a leak. The four men rigged block, tackle, and line to haul equipment aboard the rock, plus a small stove and canvas. Provisions were floated to the quarrymen. The pulling boat was forced to leave as the seas again began to rise.

The four laborers began to set up a shelter made of canvas and await the arrival of the other half of the work force. Five days later, the remaining men came aboard, bringing with them blasting powder. The powder was for leveling a foundation in the rock crest, some 120 feet above the sea.

The only shelter the party had for the first ten days was under a canvas lashed to ring bolts attached to the rock. Then the workers made a shallow niche in the north and east side of the rock, built a wooden shack, covered it with canvas, and lashed everything to ring bolts. In the words of the official report, this "gave safety, . . . but little comfort."

When work started, the plan was to first make a level bench for a derrick at the lower level of the rock then work upward to the crest to blast out the foundation of the light structure. The summit had to be dropped from 120 to 91 feet for the station. The work of blasting on such a surface proved extremely difficult. First off, there often was no place to gain a footing when drilling powder holes, nor any room to hide when the charges were set off. The workers set iron ring bolts into the walls of the rock for handholds and footholds. The quarrymen ran ropes between the rings and erected a crude wooden staging on which to work, sometimes working hundreds of feet above the churning water with the whole contraption swaying in the wind, the spray occasionally even reaching the staging.

One obstacle came from an unexpected quarter. The lower reaches of the rock had been the domain of sea lions, with some hardy ones reaching the upper level. The mammals did not take kindly to the human intruders. Some bolder sea lions attacked the workers, but the men managed to control the upper reaches and drove off their flippered attackers. Eventually, the sea lions deserted the rock.[81]

A transfer method successfully used at another difficult station, St. George Reef, California, began at Tillamook and solved the continual problem of landing supplies by boat—a heavy hawser stretched from the mast of the lighthouse tender, anchored just far enough off the rock to keep it away from the action of the waves, to highest point of the rock. This provided a strong rope bridge between the two points. A breeches buoy transferred people over the hawser. A breeches buoy looks like a pair of canvas trousers on a life ring and the ring is attached by rope to a single sheave. The sheave is placed on the hawser, and a line pulls the breeches buoy along the hawser. In effect, the hawser becomes a track. The breeches buoy made the trip to and from the rock a much safer method of bringing workers to the site. The passage from the tender to the construction site and return, however, was not without its difficult moments. As the ship rose and fell with the swells, so did the hawser upon which the breeches buoy traveled. One minute a worker would be moving through the air and the next plunged into the cold waters of the Pacific, only to spring back into the air. "The landings were frequently the basis of wagers as to how many times those in transit would be immersed. . . ."[82]

The materials for the completion of a derrick became one of the first major supply efforts. With lines, block and tackle, and wires, plus a great deal of human muscle, parts came aboard the rock and workers assembled the derrick. The derrick would have a long boom. The boom and derrick would be the principal means of bringing keepers and equipment aboard the light.

To say that life for the workers was arduous is to understate on a grand scale. For example, when the work first began, the men did not have a comfortable place to get away from the elements. Instead, they had a rude canvas shelter in the form of an A-tent. Ballantyne said, "It was rather disagreeable in our tent, it being six by sixteen feet with a horizontal ridge pole about four feet six inches from the ground. The tent, which is our only shelter, holds the ten of us. We always do our cooking on the lee side and shift with the wind direction." A wooden shack wrapped in canvas eventually replaced the tent.[83]

Most of the time the wind and seas were so strong that the workers found it easier to crawl around on the rock instead of walking. The canvas would flap loudly in the wind, causing sleepless nights, and sea spray often swept over the shelter, threat-

The "normal" seas around Tillamook Rock. (Coast Guard Museum, Northwest)

ening to wash it into the Pacific. The cook stove frequently filled with salt water, and salt water drenched the supplies.

Then came winter.

Early in January 1880, the tender could not hold station and retreated in the face of rising seas. The infamous winter rains of the Oregon coast began, along with a southwest gale. Drenched to the skin, the men continued to work, hardly able to keep their footing in the face of the tempest. Ballantyne finally gave the order to stop work and retreat to the wooden shack. The foreman ordered more ring bolts hammered into the rock and additional ropes lashed around the shed.

The storm drove salt water under the door jamb and the roof leaked. The roar of the seas was so loud that the men shouted to be heard. The rock shook as the waves hammered the crag. At two in the morning there came a crash much larger than all the other noises combined. Ballantyne ordered the men to remain within the wooden shack, while he took a storm lantern and set out to investigate. He made only one step outside the dwelling before the combined wind and seas literally hurled him back into the shack. The foreman waited for two hours and then crawled out. This time he remained two minutes, but still received a beating. For all his efforts, Ballantyne could see nothing in the storm-tossed night. All hands returned to their bunks to await the morning. Now fragments from the rock began to land on the roof. The number of rocks flung by the sea began to mount. Then the stones began to puncture the roof. The foreman ordered more canvas unfurled and stretched over the top of the wooden structure. With dawn, the workers saw the reason for the large crash: the seas had carried away the storehouse, along with the fresh water tank and most of the food.

On 18 January, a revenue cutter finally braved the seas to see whether anyone was alive on Tillamook Rock. A volunteer boat crew came dangerously close to the rock to find that everyone was alive, but living on hardtack, coffee, and bacon. The cutter remained in the area hoping to get some supplies to the workers. Four days later, the weather moderated, but the seas remained too high for small boat operations. The skipper of the cutter turned his attention to working a line to the rock. He ordered a light messenger line attached to a cask and floated to the waiting workers. This method proved futile, when the cask broke up in the seas halfway to its destination. The captain thought for a while and hit upon a different approach. Gathering up barrel staves, along with some bed sheets, someone fashioned a kite. Next, a crewman carried the kite to the crow's nest and attached a light messenger line. The crew flew the kite toward the rock. Strangely enough, this proved the device needed to pass a line to Tillamook Rock. Once the bedraggled workers had a light line, they then used that to pull a heavier hawser across the storm-tossed seas and then the much-needed supplies arrived. A lighthouse tender hove to later and landed more provisions. As more workers arrived on the rock, an improved shelter went up. A few days of mild weather helped speed progress.

After leveling the derrick site, workers then hacked a stairway through the rock to the upper elevation, and the foundation of the light was leveled. The time to land the heavy stones used in the building of the light structure was approaching. The stone used was "a fine-grained and compact basalt" and was quarried at Mount Tabor, six miles east of Portland, Oregon. The completed derrick, now equipped with a steam engine and its long boom, lifted the heavy blocks. Workers laid the cornerstone of the light on 24 June 1880. The dwelling area is 48 x 45 feet, with an adjoining 32 x 28–foot structure to hold the fog signal equipment. A 16-foot square tower rose from the middle of the dwelling to 48 feet in the air, making the beam of light 136 feet above mean sea level. Life on the rock was never easy for the workmen, but the project progressed steadily. A dramatic incident on the night of 3 January 1881 illustrated the need for the light.

The usual bad weather had set in around the rock, when the construction boss entered the dwelling and stated he saw the running lights of a ship coming toward the rock. The dim glow of red and green running lights loomed in the night. So close did the ship pass the rock, workers heard the shouted command "Hard aport!" The laborers began to kindle a large fire and lanterns were placed about to help outline the rock. The ship, which later proved to be the British bark *Lupatia,* missed the rock, but her skipper made the wrong turn and slammed into the shore of the mainland causing the loss of the entire crew.

Three weeks later, on 21 January 1881, the first-order light on Tillamook Rock officially went into operation. The light station took 575 working days to complete at an expense of one death and $123,493. The workers anxiously awaited departure, very happy to turn the site over to the lighthouse keepers.

Who coined the name "Terrible Tilly" for the Tillamook Light Station is unknown, but it proved apt. During a fierce gale in January 1883, rocks were torn loose, thrown into the air, and crashed onto the fog signal building with such force that the rocks punched twenty holes its iron roof. This was a forecast of things to come.

In December 1886, a mass of concrete filling, estimated to weigh at least half a ton, was ripped loose and flung one hundred feet above the ocean and landed near the station. Over the years, the sea seemed to try to outdo itself. In a severe storm in December 1887, the keepers reported that the seas broke continuously over the entire station, including the tower, some 133 feet above the water. On 9 December 1894, seas again breached the entire station, this time destroying thirteen lantern panes, chipping the lens, and tearing off large rocks, which were flung upon the

A keeper in a small boat prepares to be placed in the breeches buoy seen hanging from the end of the Tillamook Rock Light Station's derrick. One keeper said the ride "made a roller-coaster seem like a merry-go-round."

roofs of the dwelling and foghorn and opened the buildings to large quantities of sea water. Three years later, the new telephone cable was broken. Other damage sustained throughout the years made the station "known [around the seacoasts of the world] as the most treacherous of warning posts." The U.S. Lighthouse Board felt that the installation of a new roof of steel I-beams and concrete in 1898 would prevent further damage to the roof from rocks hurled by the seas. This would seem to be enough to stop any water from entering the station, but then came the storm of October 1934.[84]

In October 1934, Terrible Tilly felt the fury of an unusually strong southwest gale. At one point the keepers, in what must have been amazement and an awful moment of impending doom, realized everything on the rock, including the tower rising 133 feet above normal sea level, was completely under water. The lens room filled with water, and the station came close to being destroyed. As a result of the 1934 storm, Tillamook Light Station received an electric third-order lens from the Great Lakes and an iron mesh curtain to place around the glass panels in the lantern room to help prevent future damage from flying boulders. (A complete account of the keepers during the 1934 storm is related in chapter 4.)

The appropriate lead into the final chapter in the story of the Tillamook Rock Light Station is a notice of a public hearing on 1 March 1956, at Astoria, Oregon. Seventy-seven years after Superintendent H. S. Wheeler set out from Astoria to survey the rock, the U.S. Coast Guard, citing the changing nature of ocean navigation, the ability to place better aids to navigation for small boats near the rock, and the cost to maintain the facility, notified mariners of the closing of "Terrible Tilly." On 16 September 1957, Oswald Allik, the last civilian head keeper of Tillamook Rock Light, entered his feelings in his log:

> Farewell Tillamook Rock Light Station. An era has ended. With this final entry and not without sentiment, I return thee to the elements. You, one of the most notorious and yet most fascinating of the sea swept Sentinels in the world. . . . Through howling gale, thick fog and driving rain your beacon has been a star of hope and your foghorn a voice of encouragement. May the elements of nature be kind to you. . . . Keepers have come and gone; men have lived and died, but you were faithful to the end. . . . Your purpose is now only a symbol but the lives you have saved and the services you have rendered is worth[y] of the highest respect. A protector of life and property to all, may oldtimers, newcomers and travelers along the way pause from the shore in memory of your humanatarian [sic] role.[85]

Strangely enough, this is really only the opening of the final chapter of "Terrible Tilly's" waning years. Some people wanted to buy Tillamook Light. Part of the reason anyone would want to purchase "Terrible Tillie" comes from the desire of many to own a lighthouse, but knowing absolutely nothing about the light. A family who spent brief visits on the rock in the summer, arriving and departing by helicopter, bought the light. The family couldn't keep up the expenses of maintaining the building and sold their lighthouse to a young investor, who, on one approach to the light by boat, was injured, and a member of his party drowned. The best story concerns a Las Vegas group that reportedly wanted to put a gambling establishment on the rock. As humans remained off the island, wildlife began to reappear. Cormorants and murres returned to the crag in great numbers. The amount of guano made what used to be a bleak, dark rock with a white lighthouse atop it now look like a rock covered with snow. The latest, and probably the last, role for the former lighthouse began in 1980, when crews began to remodel the light into a repository for human ashes, to be known as the Eternity at Sea Columbarium.[86]

Makapuu Point

On 9 October 1888, a group of shipowners, shipmasters, and others interested in promoting maritime trade between Hawaii and the rest of the world, addressed a letter to Lorrin A. Thurston, minister of the interior, about the importance of a light at Makapuu Point.[87] The petitioners believed that a light at the point could prevent ships from running aground. Long delays were usually a part of establishing any lighthouse, and Makapuu is no exception.[88]

Thurston, in his 1890 annual report, requested a light at Makapuu Point. Superintendent W. E. Rowell, of the Department of Public Works, was the next official to get into the process. Rowell's department would be responsible for the building of the light. Thurston instructed Rowell to contact Chance Brothers and Company, of Smethwick, England, builders of lighthouse apparatus, and request costs for lamps, lens, and a lantern. The company responded in October 1890 that they thought "in your clear atmosphere, so free from fogs, a 3rd Order Revolving . . . Light with a large burner would be amply sufficient and this arrangement will also be cheaper for your stone tower."[89]

Thirteen years later, the lighthouse was still in the planning stage, but the Department of Public Works now felt the third-order light should be a fixed light, instead of a flashing one. The light would show red sectors on each side of the white to warn navigators of the reefs along the shore.

Here matters rested, for most officials believed the United States would soon be assuming responsibility for the navigational aids in Hawaii. The U.S. Lighthouse Board did take over the territory's lighthouses but with hardly enough appropriations to keep the current lights in good repair. No funds were appropriated for new construction. Eventually, the board recommended passage of H.R. Bill Number 5294 for the establishment of a lighthouse at Makapuu Point and requested an appropriation of $60,000 to build the station. In 1905, the House of Representatives took up the Makapuu light again. The bill passed on 30 June 1906, with the appropriation for $60,000.[90]

President Theodore Roosevelt issued an executive order on 12 January 1907 reserving an area of 9.82 acres for lighthouse purposes. In addition to this public

The dramatic location of the tower of the Makapuu Point Light Station, Hawaii, 420 feet above the sea. A trail to the three keepers quarters is seen stretching to the left, and the quarters can be seen just above the trail.

land, 7.906 acres were acquired on 7 August 1907 and another 11.1 acres on 29 April 1908, eventually bringing the size of the designated site to 28.826 acres.[91]

Makapuu's summit is 647 feet above the sea and consists of lava-flow layers. Between the summit and a wave-cut terrace, a site was selected for the lighthouse, about 420 feet above sea level. A trail across the steep lava rock incline was blasted, as was the shelf for the light tower.[92]

A new road ran from the main highway to the site for the keeper's quarters. There would be three keeper's dwellings, located approximately twelve hundred feet uphill from the lighthouse in a slight depression on the ridge. The houses, built of lava rock gathered from the property, had fourteen-inch-thick walls. They had white-painted trim and red-shingled roofs. Each building contained a living room, two bedrooms, a kitchen, two clothes closets, a bathroom, and a front porch.[93]

The light tower itself, because it sat so high above the sea, did not have to be tall. By October 1908, workers finished the forty-six-foot tower. The thickness of the conical tower at the base is twenty-seven inches, tapering to twenty inches at the parapet. From the base of the tower to the ventilator ball of the lantern measures approximately thirty-five feet, while the diameter of the tower's base is fourteen feet, ten and one-half inches. The foundation is of reenforced concrete block and eight feet below the surface.[94]

Makapuu's lens is special. Originally, the plan called for a third-order lens, changed upward to a second-order, and, finally, to a first-order. The U.S. Lighthouse Board decided to use a huge "hyperradiant" lens. Mentioning that the lens is huge is an understatement: the inside diameter of the optic is eight and three-quarters feet. Part of the reason for the use of this lens is that the board had purchased one and displayed it at the Chicago World's Fair in 1893, where it "was one of the most popular [exhibits] at the fair."[95] After the fair, however, the lens was stored away. Improvements in lighthouse illuminants, such as incandescent oil vapor, no longer required lenses of such size. Makapuu would have a large first-order optic probably because the board "decided that rather than purchase a new lens, they would use the hyperradiant lens they had already acquired." The lens at Makapuu is the largest lens ever used in a U.S. lighthouse.[96]

As mentioned, the inside diameter of the lens is eight and three-quarters feet, large enough for several people to stand within. It has a height of twelve feet and weighed, "as shipped with its pedestal, about fourteen tons." The keepers worked the lamps within the lens. As one historian of Hawaiian lighthouses notes, inside this giant lens "one has the weird sensation, not that the lens is so gigantic, but that one has suddenly become very small."[97]

On 1 October 1909, the Makapuu light was first displayed. The characteristics of the light were: fixed, white, with a seven and one-half-second flash and a one and one-half-second dark period. The light's beam reached out some twenty-five miles to sea. To cause the flash, the giant lens revolved on ball bearings.

An official inspection of the station in 1910 revealed that the location of the sta-

The hyperradiant lens of Makapuu Point Light Station is the largest ever used by the U.S. Lighthouse Service. The inside diameter of the lens is eight and three-quarters feet; the whole object dwarfs the man standing beside it.

tion was "healthy," while another inspection in 1911 reported that the light station was "provided with a wagon and team of mules with fittings for transportation and hauling between station and Honolulu." Apparently, life at Makapuu was pleasant. In 1925, however, tragedy struck.[98]

First Assistant Keeper Alexander D. Toomey and Second Assistant Keeper John Kaohimaunu were changing watches, when Toomey suggested that Kaohimaunu fill the alcohol lighter, located in the light's bottom-floor service room. The alcohol was used for heating the oil-vapor lamp. Most of the alcohol from a tank was drawn off, with some dripping on the floor. For some unknown reason, Toomey struck a match and the room exploded. Kaohimaunu, standing near the door, received serious burns, but managed to stagger out of the room. Toomey caught the brunt of the explosion and was "charred black and crinkled," but somehow managed to reach his quarters.[99]

The head keeper had to transport both the first and second keepers to a hospital. Toomey "refused to permit his wife to accompany him and insisted that she remain at the lighthouse, which would necessarily be without a keeper for a considerable time." Toomey did not survive his injuries. His wife, Minnie Ululani Ua, gave birth to a daughter, Violet, soon after her husband's death. Three months later she also died, leaving the infant and a daughter of fifteen.[100]

Two years after Keeper Toomey's tragic accident, the Makapuu light received the latest wonder in aids to navigation: a radio beacon. The beacon was the first in the Hawaiian Islands. Two 80-foot galvanized steel towers were built on the promontory behind the light to carry the 180-foot antenna. A wooden structure housed the equipment for the beacon.[101]

By this time, Makapuu had its own electrical generating plant. Electricity also meant a new source of power for the light. Electricity and the radio beacon meant, too, that the duties of the keepers changed. Now the lighthouse service employees spent most of their time monitoring the radio beacon, rather than the light.

The discontinuation of the radio beacon on 1 November 1973 pointed to the eventual closure of the light. Another important turning point came on 27 October 1963, when Joe Pestrella, a civilian light keeper, retired from Makapuu light after twenty-four years of service. He was the last civilian light keeper in Hawaii. On 4 January 1974, U.S. Coast Guard personnel left Makapuu light and placed it in automated status.[102]

Although closed, the keeper's quarters for a brief period in 1974–1975 was used for witness protection in a federal tax evasion trial. Later, vandals broke into the quarters and ransacked the buildings. In 1984, vandals shot a hole through the hyperradial lens. The damage to the lantern room's window was repaired, but the lens could not be fixed. As one author pointed out, automation has "simplified and economized the operation of Makapuu Lighthouse, but it has also left it vulnerable." In 1987, the U.S. Coast Guard declared forty acres of the Makapuu Point light as surplus, including the land that held the keeper's quarters. This surplus land

The hyperradiant lens in the lantern room of Makapuu Point Light Station gives an idea of the light's huge size. Note that one of the visitors is taking advantage of the hand holds around the panes of the lantern room for keepers to grasp in stormy weather or while they clean the windows. The ball vent above the lantern room is also clearly visible.

became a part of a land-ownership debate with native Hawaiians. Soon after the settlement, the state tore down the buildings.[103]

The area around the former light station is now a state park, with the tower and about five thousand square feet still belonging to the U.S. Coast Guard as an active aid to navigation. The state of Hawaii opened the narrow winding access road to the public for hiking. In May 1996, Coast Guard Auxiliary Flotilla 23 planned to "host escorted tours to the lighthouse monthly." McKinnon Simpson, an historian of the Honolulu-based Hawaii Maritime Center, noted that the hike to the light tower and the view "are both breath-taking," but visitors should be aware that the hike is "long, and mostly uphill. There are also sheer drop-offs with no guardrails or fencing. And a broiling sun."[104]

Keepers and Their Lonely World

TODAY, PEOPLE VOLUNTEER to become "keepers" at lighthouses that are now saved by lighthouse organizations, historical societies, and state and national parks. This is a far different situation than what many keepers recorded during the years of the U.S. Lighthouse Service. The words they most used to describe their lives were *loneliness* and *monotony*. One northern Michigan keeper lamented his "lonely station." A keeper at Point Reyes, California, wrote, "O Solitude where are the charms that Sages have seen in thy face. Better dwell in the midst of alarms than reign in this terrible place."[1] Yet, the service usually did not lack for want of employees. Who kept the lights? What was it like to be a keeper of a light station in the U.S. Lighthouse Service?[2]

The public's perception of the lighthouse keeper is that of a competent, kindly man. He is largely seen as a favorite uncle, puttering around a lighthouse, telling sea stories, and worrying about the dark. In short, the image of the keeper has worn well with the public. Until near the final decades of the nineteenth century, however, the historical record depicts a different image. George Worthylake became the first keeper of Boston Light Station in 1716 and, thus, the first keeper in the Lighthouse Service. The act that established Boston light also explained the duties of Keeper Worthylake. The eighteenth-century guidance, with allowances given for changing technology, is applicable for at least the next three centuries. Worthylake's instructions said he should " . . . carefully and diligently attend to this Duty at all times in kindling the Lights from Sun-setting to Sun-rising, and placing them so they may be most seen by vessels coming in or going out." For his labors, Keeper Worthylake was to receive a salary of fifty pounds a year.[3]

For well more than thirty-three years after the appointment of Keeper Worthylake, the quality of the keepers in the United States, according to the major investigation of the service in 1851, "was very various."[4] Collectors of customs usually appointed

A keeper winds the weights until they are near the top of the tower. The falling weights made possible the movement of the lens. (Evanston Historical Society, courtesy U.S. Lighthouse Society)

keepers. The collectors owed their positions to politicians and appointed applicants for lighthouse keepers from the political party in power. Other politicians saw the service as a reward to cronies. If the political climate changed, so did the collector, or political patron, and, in all likelihood, so did the keeper. James Miller, in an 1845 letter to the secretary of the treasury, wrote that he had fought against the British and "was wounded & in consequence lost my leg. . . . Since which time I have managed to get along with one leg. And am now the present incumbent or keeper of the lighthouse on Stony Point on the Hudson River. From which (as *your honour* is fully aware) strenuous efforts are making to have me removed; and the only charge *or the most* tangible one is my having voted for *Gen. Harrison* in 1840. . . ." An old salt wrote that he had "no occupation but a sailor having followed the sea since 12 years of age . . . I think I ought to be entitled to the patronage of the government in preference to those lazy land-lubbers who keep the lighthouses and light boats in Long Island Sound. . . ." Joshua Lane, of the Portland Light Station, Barcelona, New York, slyly addressed his request for a continuation of duty to First Lady Polk.[5] An organization whose purpose was to guide sailors to safety became riddled with many who sought only a reward for choosing the right party. Nowhere is there any thought of proper qualifications for the position of keeper or in keeping permanent employees. Not every keeper was an inept political appointee. There were enough, however, to throw doubt upon the workers of a service whose single duty could literally cause the deaths of many sailors and passengers.

Just where the early lighthouse keepers learned their trade remains unknown. According to one source, an instruction sheet posted in the lantern room supplied needed information for the keeper. The 1851 investigation of the service, however, revealed that most lighthouses were without this notice. The new keeper may have received training from the former keeper or from the collector of customs. In any case, poor instructions and training are two good reasons for the poor quality of the early lights in this country.[6]

In general, the keepers of the early lights ran the gamut from terrible to conscientious. The 1838 investigation of the lighthouses, however, suggested too many keepers went about their duties lackadaisically. The dividing line between a service of political appointees, with haphazard accomplishments, and a professional service is, again, with the establishment of the U.S. Lighthouse Board in 1852.

The U.S. Lighthouse Board recognized that politics had no place in a service that others depended upon for safety. Recognizing and then solving the problem, of course, are two very different matters. The U.S. Lighthouse Board tried to prevent political appointments, but the interest of politicians made the path to a career-oriented service a slow process. Harbingers of success began with the passage of the Civil Service Reform Acts of 1871 and 1883. President Grover Cleveland's executive order of 6 May 1896, which placed keepers' jobs in the classified civil service positions, marks the solution of the dilemma.[7]

Prior to the Civil Service Reform Acts, the U.S. Lighthouse Board laid the foundations for a career-oriented service with two important decisions focused on the keeper. A first step centered on the training of the keeper. To offset the lack of training and instruction of the early keepers, the U.S. Lighthouse Board began the process of setting down written instructions in the minutest detail. Lighthouses began to receive such publications as *Instructions and Directions to Guide Light-house Keepers and Others Belonging to the Light-House Establishment* and others that provided a keeper with the information needed to operate a lighthouse. The instructions presumed an employee could read. Therefore, the U.S. Lighthouse Board's first two instructions concerning new employees stated an applicant had to be able to read and to be at least eighteen years old.[8]

To say that the instructions spelled out everything a keeper needed to know about keeping a light is understatement to the extreme. The instructions, for example, demanded that only "substantial and wholesome food . . . would be provided at the station" and "articles known as luxurious are forbidden to be provided" by the service.[9]

Interestingly, the U.S. Lighthouse Board even printed detailed instructions on how to administer the U.S. Lighthouse Board. One could set up and successfully run a U.S. Lighthouse Board today with these printed guidelines. For example, the board was to "put the corrected Light-house List in the hands of the printer as soon as possible after the 1st day of January of each year, and correct proof-sheets as they are received." The board was to distribute "new Light-house Lists as soon as ready according to list" and distribute "Buoy lists." Then there was the daily routine that required attention in the office of the board. Leading off a list of thirty-six items to accomplish was the admonition that "all letters received to be endorsed and registered immediately"; following was the instruction that is not always obeyed today: "The mail matter to be attended to, as far as possible, on the same day that it is received."[10]

The second major decision of the U.S. Lighthouse Board in an effort to improve keepers comes under the rubric of performance of duty. Most contemporary followers of the lighthouses of this country note the large difference in how well the keepers discharged their duties after the U.S. Lighthouse Board took control. The reason, in part, for this change is that the instructions printed by the board led to better-trained keepers. The U.S. Lighthouse Board also set high standards and enforced them with fair but swift justice. Coupled with rigorous inspections, good training, high standards, and enforcement of the regulations and standards finally led to a professional service.[11]

An added fillip on the path to a professional lighthouse service came in 1884 when the U.S. Lighthouse Board introduced a uniform for all male lighthouse keepers and officers of lightships and lighthouse tenders. The dress uniform consisted of "dark indigo jersey or flannel." A double-breasted coat sported a double row of five metal buttons running down the front. A yellow metal lighthouse badge sat just above the visor on the cap. The board issued the first uniforms in 1885.[12]

Keeper Edward H. Schneider gazes from his light tower on Alcatraz. His view encompasses San Francisco Bay and the federal prison at Alcatraz Island, circa 1954. (Each stripe, incidentally, near the left-hand cuff of his uniform jacket represents four years of service.)

The old handwritten personnel ledgers of the lighthouse service now resting in the National Archives are witness to the change in keepers. In the 1850s and 1860s, lighthouse service clerks recorded "resigned" or "dismissed" in the column labeled "reasons for leaving" an individual station. Entries of "transferred" and "promoted" appear more often after the 1860s and by the turn of the century "transferred" and "promoted" are the standard entries in describing why a keeper left a station.[13] By the beginning of the twentieth century, the U.S. Lighthouse Board could proudly point to the fact that they had taken a group of employees, once charitably described as "various," and molded them into a dedicated and professional corps of keepers. These newly made professionals, in turn, produced the finest lights in the world. Succinctly and correctly, the lighthouse historian F. Ross Holland writes that probably "a chief cause for the board's success was the military background of its members and their sense of military justice: fair, stern, and swift."[14]

WHAT WERE THE DUTIES of a lighthouse keeper? Quite simply, the primary duty of the lighthouse keeper was to keep a good light for the mariner. This deceptively easy task required a good keeper to spend many working hours at the light, both day and night, and usually 365 days a year. When the spider lamps were the primary lamps in lighthouses, the keeper's labors included keeping the wicks lighted, the oil reservoirs full, and the panes of glass cleaned of the accumulated smoke from the lamps. After the replacement of the spider lamps with Argand lamps and parabolic reflectors, the polishing of the reflectors required additional hours of toilsome labor to keep the reflectors bright. The constant attention to cleaning and polishing brass led some wag to produce a long poem on the subject, part of which was entered into the log of the Point Reyes Light Station:

> Oh, what is the bane of a lightkeeper's life,
>   That causes him worry and struggles and strife,
> That makes him use cuss words and beat up his wife,
>   It's brasswork.[15]

Wicks needed to be trimmed each day to remove the residue of burning. (One of the most important duties of keepers, of course, is making sure the light burned brightly. In the days of lamps, the best and brightest lights were attained by keeping the wicks of the lamps well trimmed. The attention devoted to wicks by keepers quickly led to all keepers earning the sobriquet "wickie.")

The U.S. Lighthouse Board's written instructions divided the work schedule in lighthouses of two or more keepers into two departments. The assistant keeper took charge of the first department, from sunset to around midnight. The assistant would remain in the watchroom until needed. The first department's duties included cleaning and polishing the lens. Next came cleaning and filling the lamp, along with dusting the framework of the apparatus. The all-important task of trimming the wicks of the lens lamp, and putting new ones in, if required, was yet another assignment. The head keeper took up the duties of the second department and these duties centered on cleaning. This keeper cleaned the copper and brass fixtures of the apparatus as well as the utensils used in the lantern and watchroom; cleaned the walls, floors, and galleries of the lantern; and swept and dusted the tower stairways, landing, doors, windows, window recesses, and passageways from the lantern to the oil storage area. The keeper was reminded not to clean the floor with "any material by which dust may be produced." The instructions even informed the keepers to wear linen aprons to prevent their clothing from scratching the lens. After extinguishing the light, a curtain was drawn around the lantern room to protect the lens. All work required to have the light ready for the first department had to be completed by ten o'clock in the morning. Instructions also went on to spell out the longer-range routines for the light. The book instructed the keepers when to wash the lens (every two months), to polish the lens annually with rouge, and other preventive maintenance duties.[16]

Most light stations on land contained the tower, keeper's quarters, perhaps a fog signal building, a building for storing oil, a boat house, wharf or landing area, and sometimes a few additional miscellaneous structures. After finishing the normal routine and long-term maintenance of the light, keepers then undertook work on the

George Esterbrook, third assistant keeper at Cape Disappointment Light Station in Washington, escaped from the upper gallery on a lightning rod cable. Apparently, the stairway to the lower gallery was not in place at the time Esterbrook found himself locked out of the tower during a storm.

"Oh, what is the bane of a lightkeeper's life,
That causes him worry and struggles and strife? . . .
It's Brasswork.
I dig, scrub and polish, and work with a might,
And just when I get it all shining and bright,
In comes the fog like a thief in the night.
Goodbye Brasswork."

station, such as painting and ordinary repairs. The 1881 instructions, on page twenty-eight, detail a recipe for whitewash that "by experience . . . [works] on wood, brick, and stone, nearly as well as oil paint, and is much cheaper. . . ."[17] Painting of the light tower by the wickies for some reason fell under routine work and was the largest task. Stations whose principle keepers were women were excused from this chore. Work forces under the control of district engineers handled any large-scale work on the station site. Once again, the instructions told a keeper everything needed to keep a good light and how to keep the station inspection ready. If a question arose regarding how to complete a task, the keeper scanned the instruction book's index until he found the proper page and then looked up the answer. The wickie had only to read and have a normal amount of intelligence to run the lighthouse.

Perusing the instructions, one can see just how regimented the employees of the service actually were and also the service's strong emphasis on cleanliness. Of course, the lanterns and lens, plus the windows in the lantern room, had to be clean to produce a good light, but sometimes a few employees went to great lengths in pursuing the service's decrees on neatness. One keeper on Buzzard's Bay, Massachusetts, for example, asked the inspector to put felt slippers over his shoes while ascending the light to keep the stairs clean.[18]

The simple routine of cleaning windows could at times be dangerous. Standing high on a light tower could be hazardous in heavy winds. Many light stations had hand grips along the vertical window frames to prevent keepers from losing their footing during storms.[19] Third Assistant Keeper George Esterbrook, of the Cape Disappointment light, had a harrowing experience with the wind after he relieved the watch. Esterbrook went to the lantern room and then out onto the gallery to start cleaning the windows. A storm raged along the Washington coast as usual. No sooner had the keeper stepped out onto the wind-and-rain-lashed gallery than he heard the door slam heavily. Esterbrook checked the door to find it locked shut. He was now trapped on a dangerously windy and wet gallery some fifty-three feet in the air with no way to call for help. Esterbrook recalled a lightning rod cable that ran down the side of the tower. Below the main gallery that the keeper was trapped upon, Cape Disappointment light had another gallery. Esterbrook managed to grab the cable and inched himself over the rail and started down. The wind was so strong that one gust swung his body outward until it was close to parallel to the tower. The wind then slackened, and Esterbrook slammed into the building. The keeper lost his grip, but managed to land on the lower gallery. Esterbrook found an open door and made it safely inside.[20]

Within the routine of keeping a good, bright light was a problem that could add a little change to the ordinary work day. Birds constantly struck lighthouses. On Maine's Saddleback Ridge light in 1927, the keepers were sitting in the kitchen talking about World War I, "when bang, bang, something came against the window panes." The keepers "thought another war had started. . . ." At least 124 ducks had struck the light.[21] In 1902, Keeper Bernard J. Bretherton, wrote in the *Osprey* that most birds struck lighthouses in thick weather and not during normal darkness. Bretherton felt that the black paint pattern outside the lantern room, balcony, and below the lantern room became invisible in foggy weather. The birds see only the light, which they take "for the moon, a star, or other harmless heavenly body." The birds steer toward the light and at the last minute the bird may see the windows, swerve or dive to miss them and, not seeing the black portions of the building, will run into the structure. According to the keeper, at least "ninety per cent of the birds are killed" hitting this area of the light tower. Larger birds, such as ducks or geese, could break the glass of the lantern room. Bretherton also noted that bird collisions are more apt to happen at new light stations, and the numbers decrease as the years pass. To prevent damage to the lantern room glass, the lighthouse service used iron screens during periods of heavy bird migration.[22]

Charles Separdson, Barney Lokken, and Oscar Linberg, the 1927 keepers of the isolated light station at Scotch Cap, Alaska. (USCG, courtesy U.S. Lighthouse Society)

MANY PEOPLE IN OUR MODERN WORLD see the isolation of lighthouses as the ideal retreat from the everyday pressures of life. There is, however, another side to the isolation. One keeper said it very succinctly: "The trouble with our lives is that we have too much time to think."[23] The combination of isolation and monotonous routine can lead some people to do or imagine strange things. Alaska had very isolated lights. Commissioner of Lighthouses George R. Putnam said that the Cape Sarichef Light Station, located on the Bering Sea–side of Unimak Pass in the Aleutian Islands, was the most isolated station in the service. Keepers spent three years on duty and received one year off. The sea legend of Lee Harpole has persisted throughout the years.

Ships did not use Unimak Pass in the winter because of sea ice in the Bering Sea, and the Cape Sarichef Light Station remained extinguished during the winter season. The nearest inhabitants were at the Scotch Cap Light Station on the Pacific side of the pass, seventeen miles away. A small trail over the rough Aleutian terrain connected the two lighthouses, and it was not unusual that keepers from one station or the other would visit each other throughout the winter period. A small cabin sat at about the halfway point so the keepers would have a place to rest and seek shelter in case of a sudden storm. The most hazardous part of the trip was the crossing of many ice-cold, rushing streams.

Harpole decided to visit his neighbors and had already passed the middle point cabin when he came upon another stream. The keeper's strategy for crossing the frigid water was to strip down to his underwear, tie his outer clothing into a bundle and throw it from rock to rock as he gingerly picked his way across the water. On one throw Harpole's aim was off, and the bundle of clothing went rushing down stream. The closest location he could obtain clothing was the Scotch Cap Light Station, eight miles away. To add to the drama, it began to snow. Incredibly, the keeper ran the distance wearing only a stocking cap and underwear and carrying a rifle. The expressions on the faces of the keepers at Scotch Cap must have been incredulous when they saw Harpole approaching through a snow storm wearing only his underwear. According to one author, it was "three days before Harpole's lowered body temperature returned to normal."[24]

Isolation can also affect the mind. At the isolated Tillamook Rock Light Station, a keeper was jolted awake with the sound of a single footstep in his room. The wickie knew one man was on watch and everyone else was in bed. Another step. Then something seemed to brush his face, followed by a light touch to his throat.

The keeper imagined a "knife" near his throat. He leaped up and started across the room to the light switch, flaying about in the darkness to ward off his attacker. His foot struck something, and the terrified keeper tripped and sprawled upon the floor. He jumped up and managed to get to the light switch. The light revealed a large, stunned goose waddling about the room. Apparently the goose had somehow managed to get into the room's porthole. The keeper's panic came about when the "bird leaped upon the bedside chair and its wing flapped across the keeper's throat."[25]

Isolation also can affect what are now known as interpersonal relationships. At another Alaska light, an inspector found two keepers not speaking to each other because "one liked the potatoes fried, while the other wanted them mashed."[26] A storm at California's isolated St. George Reef Light Station stranded the crew for a month. The normally sociable crew at the end of this period "ceased speaking to each other." By the end of seven weeks, meals "were eaten cold with the men facing away from each other." The storm finally broke after fifty-nine days, and the men "became friends again."[27]

Although some of the results of isolation can be fairly amusing to those who have not experienced being cut off from outside communication, there is a serious side of isolation. A keeper of the Cape Sarichef Light Station went insane when he thought the ghosts of the Aleut hunters killed by Russian hunters were about him.[28] The wife of a Cape Disappointment keeper committed suicide by leaping from a cliff. Further northward along Washington's coastline, at Cape Flattery, "one despondent attendant who hated his isolated role, attempted suicide by leaping into the sea." The keeper survived the leap. Isolation also meant that keepers and their families very well could die before obtaining medical attention. One author notes that at least fourteen "government employees have succumbed" at Cape Flattery. Because transportation could not be arranged quickly enough, a child of one of the keepers at the Farallon Islands station died.[29]

Despite the good press that most keepers have received from writers, not every person proved to be a good employee. There were cases of alcoholism, insubordination, and violence. The light station at Point Reyes, California, is an example. Keeper John C. Ryan took over the station in January 1888, found Point Reyes badly maintained, and started a strong work program. Ryan apparently pushed too hard, for a year later "one assistant went 'crazy'"; Ryan's dismissal followed shortly thereafter. Another keeper at Point Reyes, suffering from alcoholism, supposedly even sampled the "alcohol furnished for cleaning lamps." In 1876, the two assistant keepers to William Wadsworth threatened Wadsworth with "violent language when he tried to put them to work on road repair." Third Assistant Keeper J. D. Parker compiled a record of neglect and insubordination. For example, at one time he shut down the fog signal and reported clear weather, even though Point Reyes was shrouded in fog. All this is not to single out Point Reyes as a poor light station, rather it does point out that light stations were crewed by human beings and this means they were subject to all the frailties of being human.[30]

SIMPLY ARRIVING OR DEPARTING some lighthouses presented a great deal of danger. The Farallon Islands outside of San Francisco, for example, was a dangerous station because of the difficulty in transporting supplies and people to the location. "Boat day" at the Farallons was the day when supplies and relief keepers came aboard the island. A derrick hoisted everything onto the island, and once the supplies were delivered, a special rail car transported the material over some 3,559 feet of track. (Patsy the mule towed the car. The mule would hide whenever she heard the whistle of the tender.[31])

The transport of people from the tender to the Farallons, and vice versa, by derrick evolved over the years. At one time the person making the journey to the island sat in a regular wooden chair, and the derrick lifted the chair. Then people were placed in a "basket," which meant standing inside a boxlike structure. Later a "Billy Pugh net" (a circular, cagelike affair with ropes and a platform for standing upon) was used. Another method used the derrick to hoist a boat with everyone in it and place the boat in the water; the boat would then go to the tender, swap keepers and supplies, and repeat the process in reverse.[32]

Getting to and from a light station could be hazardous. The Eighteenth District Inspector lands at St. George Reef, 30 March 1916. A modern-day boatswain's mate remarked that when landing some-one at the light station from a boat "you had about two seconds to get them on the light or you could be in trouble."

The best description of coming aboard the difficult Tillamook Rock Light Station is by author Jim Gibbs, who served a tour as a U.S. Coast Guardsman aboard the rock. Gibbs, brought to the vicinity of the rock in a small boat, heard a grinding sound and soon saw a long boom swinging out over the boat. A cable slowly lowered to the boat. Then the boatswain's mate on the lifeboat handed Gibbs a breeches buoy and told him to get into it, stand on the bow of the lifeboat, and hook on to the cable. The boatswain's mate signaled the rock.

I felt the ropes of the breeches buoy tighten. Over the side I went, floundering in the coldest water this side of solid ice. I thought I had signed my death warrant, when with a terrific jerk I ascended from the depths like a hooked fish. Up and up I went dangling. . . . For a moment I thought I had sprouted wings, but quickly realized that my survival was dependent on the flimsy contraption which held me between water and sky. The lifeboat below dwindled to the size of a peanut shell. Strong gusts hampered the turning of the boom, as I sat there rocking, while the chill wind made ice sheets of my wet clothes. When the winch finally gathered enough power to swing the boom I breathed easier, but kept wondering why amusement park promoters had never given this invention a whirl; it would have made a roller-coaster seem like a merry-go-round.[33]

Keepers going away from the light for their allotted time off faced the same type of ride in reverse. Gibbs implied that many keepers coming to the station received a dunking as a form of amusement for the boom operator.

IN OUR MODERN WORLD, it is perhaps difficult to understand what families endured on some of the lights. Keeper Eldon Small and his wife, Connie, spent twenty-nine years in Maine lighthouses. Connie Small recalled that it was not until 1946 that they were assigned to a light that had electricity. "Eldon went wild. The week we moved in, he bought an electric toaster, an electric refrigerator, an electric coffee pot. . . ."

Connie Small also related that "Cooped up on a rock together for months on end, seeing no one else, . . . we'd some times get mad at each other," then "Eldon would stalk off to be by himself down in the engine house. Or I'd go stand alone on a ledge by the sea for a couple of hours." Eventually, what helped the couple was good communication, "a good airing out of our differences, talking about the things that griped us, was our secret to being able to live together happily, alone on a rock for months on end."[34]

To help keepers and their families overcome boredom and to improve their larder, the service encouraged the cultivation of gardens. The light station at the north end of Grand Island, Michigan, contained a "600 square foot garden, enclosed by a low stave fence."[35] Keeper Frederick W. Boesler, Sr., of the Au Sable Light Station along Lake Superior's coastline, noted in his log that he had "grafted 24 fruit trees, 12 of cherry and 12 of apples."[36] The soil at many stations prevented farming, but some people tried no matter how poor the earth. At Maine's rocky Boon Island, for example, keepers hauled dirt in barrels and boxes to start gardens, all the while knowing that the first hard storm would cause the sea to wash away their efforts. During World War I, Keeper L. A. Borches, of the Turn Point Light Station, harvested the sea. Borches canned fish and caviar and received praise from Herbert Hoover, at the time head of the U.S. Food Administration.[37]

Another means of fighting the long hours of isolation were the portable libraries introduced in 1876. The libraries contained about forty to fifty books and magazines on history and science as well as poetry. To better understand the reading habits of their employees, the service had keepers indicate which books they read from the library. The libraries changed at the time of quarterly inspections. By 1912, the service reported at least 351 libraries in circulation among the stations.[38]

Perhaps isolation, or low funds, accounts for a trait of penny-pinching among lighthouse service employees. There seemed to be an unwritten motto of "use it up, make it do, and don't throw it away." This trait seems especially strong among the officers who visited the stations. Harry W. Rhodes, superintendent of the Eighteenth Lighthouse District from 1912 to 1939, is a case in point. Rhodes, on a visit to the Point Reyes Light Station, noticed a discarded, worn-out broom. Rhodes became indignant. "Why, this could still be used for scraping whitewash off a fence. Put it back in service." Before he left the station, Rhodes gave orders to the keepers to file old screwdrivers back into shape.[39]

KEEPERS AND THEIR FAMILIES faced both boredom and danger. Some of the danger came from sickness while at an isolated location and some came from accidents in everyday work, such as going to and from a particularly difficult station. The greatest danger came from the elements. A hurricane on 27 September 1906 caused the destruction of the light station at Horn Island, Mississippi, killing the keeper, his wife, and daughter. The first light structure at Minot's Ledge (en route to Boston, Massachusetts) collapsed in a heavy storm in 1851, killing two assistant keepers. On Maine's Boon Island, storms tossed across the island boulders weighing hundreds of pounds. At Tillamook Rock Light Station, storms seemed to take exception to the light on the isolated rock and never so much as in October 1934.[40]

On the morning of 21 October 1934, Keeper Henry Jenkins turned in after his night watch. When he went to sleep, the light station was in the grip of a southwest gale with a "hissing curtain of water." Jenkins was wrenched awake trying to catch his breath and found his mouth full of sea water. The keeper thought the station had been swept out to sea. The water receded, and he discovered himself in the closet of his room. Another wave crashed through the window. Sea water was waist deep in the living quarters. Making his way to the galley area, Jenkins saw the iron stove pushed from its foundations.[41]

By ten in the morning, the mountainous seas continued their crashing down upon the station, and huge boulders could be heard rolling across the concrete-and-iron roof. Head Keeper William Hill shouted for all the keepers to go to the lantern room and bolt emergency storm panels to protect the light mechanism in case anything crashed through the glass panels. Before the keepers could reach the stairwell, however, the entire rock shuddered violently as a twenty-five-ton section of Tillamook Rock crashed into the sea, followed by a sixty-pound boulder that crashed through the glass paneling. Tons of water, broken glass, rocks, dead fish, and seaweed came rushing down the tower, flooding the interior of the living quarters and "forcing the men to climb the steel roof supports to keep their heads above water." With a sense of immediate doom, the four keepers realized that "Tillamook Light Station, including the . . . tower 133 feet above the sea, was under water." The surge passed, but the men knew there would be yet another, as the station "shuddered and groaned."

Expecting to die at any moment, between surges the keepers fought their way in waist deep water to the light tower's stairway, knowing emergency panels must be bolted into place. If not, the station was doomed. The keepers knew, however, the light must somehow be shown to protect any mariners who may be fighting their way to safety. They struggled up to the lantern room. There they found sixteen panels smashed by the boulder, along with the lamp and lens. Struggling with the panels, "submerged at times to their necks before the rush of water . . . [,] the in rushing seas brought fragments of rock and glass, and even small fish." Assistant Keeper Hugo Hanson's hand was "badly cut with flying glass."[42]

All that night and the next day, the four men worked without food or rest securing the panels and rigging an emergency light. On the second night, a feeble, steady light rather than the strong, flashing beacon shone from the rock; at least mariners would have some warning in the howling storm, and the light tower was no longer a conduit to the sea. The next step was to let headquarters know about the problems.

The U.S. Lighthouse Service knew nothing of the plight of the four keepers on Tillamook Rock. Keeper Jenkins was an amateur radio operator and attempted to make something from the useless telephone and an old battery-operated receiver. Jenkins jury-rigged a transmitter and began to transmit Morse Code, which was picked up by another amateur who lived on the coast in Seaside, Oregon, almost directly across from the light. The radio operator called the headquarters of the 17th Lighthouse District, who dispatched the tender *Manzanita* to the rock. It was not until 27 October, however, that the seas abated enough to allow inspection and work parties, along with fresh food and water, to come aboard the station.

THE PRIMARY DUTY of lighthouse keepers was the displaying of the light. Keepers, however, were also lifesavers. The location of lighthouses made them ideally suited for helping those in distress. Keeper's deeds are recorded in the official reports of the service and in the reports of the U.S. Life-Saving Service. The accounts detail everything from simply towing in a boat to life-threatening rescues. The keeper with the nation's highest achievement in medals must be Marcus Hanna. Hanna received the United States' highest military award for valor in the Civil War— the Medal of Honor—and after the war, while keeper of the Cape Elizabeth Light Station in Maine, he was awarded the Gold Life Saving Medal, the nation's highest civilian award for lifesaving. The Gold Life Saving Medal was awarded to Hanna for his rescue of the sailors of the schooner *Australia* on 28 January 1885.

The *Australia* departed Booth Bay, Maine, en route to Boston on 27 January 1885. The schooner ran into a "furious gale" of wind and snow. The skipper tried to seek shelter at Portland, Maine, but the ship was driven onto the rocks at Cape Elizabeth. The *Australia* struck around eight in the morning. The sea "poured over her in a perfect deluge," sweeping everything off the decks, including the pilot house. The sailors had just enough time to take to the rigging. The captain, however, was swept from his perch to his death a few minutes later, leaving two sailors, "drenched

Frederick T. Hatch—shown in his U.S. Lighthouse Service uniform—was one of only three men to receive two Gold Life Saving Medals during 1876–1915 for rescues at sea. Hatch earned the first award in 1883 while serving in the U.S. Life-Saving Service at Cleveland, Ohio, and the second (represented by a gold bar in lieu of a second full medal) for saving a woman in 1890 while he was lighthouse keeper at Cleveland. Charles Nordhoff paid tribute to Hatch with this etching, which he called "Frederick T. Hatch: The Only Recipient of the Gold Bar for Heroism."

to the skin," huddling in the rigging in temperatures of minus ten below zero.

Marcus Hanna had just laid down for a nap after being relieved from his watch. His wife, looking out the window, spotted the masts of the *Australia* and yelled to her husband. Marcus grabbed his coat, slipped into his boots, and ran to the shore.

He signaled to his assistant, Hiram Staples who was in the fog signal building, to follow, and the two keepers were soon at the scene of the wreck. Hanna knew it would be impossible to launch a boat. He returned to the fog signal building for an axe and then ran to the boat house for a line. The boat house door was blocked by snow, so Hanna ran back to the fog signal and told Staples to bring a shovel. Soon, Hanna had a line and rushed back to the scene. In the meanwhile, Hanna's wife had alerted the other families at the light of the wreck, but only one person, a boy of fifteen, was able to come to the wreck site.

At the scene, Hanna weighted one end of the line with a piece of metal that he had brought and worked his way down the slippery, ice-covered rocks almost to the surf. The U.S. Life-Saving Service said that his position placed him in "almost as great peril as that of the two men in the rigging of the schooner." At this dangerous location, Marcus took the weighted line and heaved it toward the *Australia*. It failed to reach the hapless sailors. Again and again Hanna heaved the line.

Hanna now became forced to retreat to the top of the rocks to restore some warmth to his limbs. At just about this time, a large wave took the schooner and threw it beam end onto the rocks, thus placing the sailors in even greater danger. Hanna, seeing this greater danger, again worked his way down the rocks. At last, the thrown line reached the *Australia*. One of the sailors made the line fast around his body. Hanna went to the top of the bank again and found no one there to help him. He yelled for help. The sailor, almost ready to give up, leaped into the breakers, and Hanna pulled him to safety. Hanna loosened the line from the first sailor and heaved it across to the remaining tar and the process was repeated. Just as the last man came ashore, Staples and the others came rushing up and helped hurry the sailors and Hanna to warmth. Adding to all the danger was the fact that Marcus Hanna "had been ill for a week or more, and it was only through the exercise of a most determined will that he was able to withstand the hardship and exposure. . . ."[43]

ALTHOUGH KEEPERS AND THEIR FAMILIES faced a great amount of boredom, isolation, and danger, most of the people who remained long-term employees developed a love of serving on the light stations. On the Farallons, for example, Keeper John Kunder "seldom went ashore, even for his leave."[44] The very isolated and dangerous Tillamook Rock Light Station seemed to exert a strange pull upon those who served on it. For his long and faithful service at the isolated station, one keeper received as a reward the assignment to help care for the U.S. Lighthouse Service's exhibit at the Pacific Exposition in San Francisco in 1898. After a week among the crowds, the keeper pleaded to return to his rock. "No more of them noisy wise cracking crowds for me, I'll live here until I die."[45] Robert Gerloff actually hated to go ashore, and it was rumored that he had spent a five-year stretch on the rock without relief. When it came time for his retirement, Gerloff pleaded with the service to allow him to stay aboard Tillamook Rock Light Station as a paying guest, but his request was denied and the old keeper was forced to leave the rock.[46] A 1952 U.S. Coast Guard report said there were two civilian keepers on the rock with two U.S. Coast Guardsmen. The head civilian keeper, George H. Wheeler, had twenty years of service on the rock. According to the report, he had been "offered transfer to various shore stations with family quarters, but he prefers to remain where he is." Oswald Allik, the assistant civilian keeper, had fourteen years on the rock and did not want a transfer, "although he has been offered various shore stations with promotion to keeper."[47] Ted Pedersen spent five years at the extremely isolated Cape Sarichef Light Station, Alaska, but recalled, "Cape Sarichef was home to me, a place

I wish I had never left."[48] Lighthouse keeping sometimes spanned several generations. Keepers James E. and George F. Sheridan's parents, for example, kept lights
on Lake Michigan. Their grandparents were keepers at South Manitou Island, as
were their cousins.[49]

There is the interesting situation of a keeper, who Commissioner George Putnam
called Malone, and the isolated Isle Royale Light Station in Lake Superior. Malone,
who was single, helped in the construction of the light. When finished in 1875,
Malone applied for the job of keeper. The inspector informed Malone that the station was for married couples, whereupon Malone went off to the mainland and married. He received the position. Malone raised a large family of twelve on the isolated
island. A new inspector was transferred into the district about every two years. The
Malones "adopted a custom of naming a child after every Inspector, but the system broke down when the exigencies of the Spanish American War caused three different Inspectors to be detailed in one year."[50]

When the U.S. Coast Guard took over the lights in 1939, very few civilian
employees of the U.S. Lighthouse Service chose the option of becoming military
personnel. Keepers probably felt they ran a risk of being removed from their light
stations if they assumed military status. This feeling would be proven correct during World War II. The keepers' refusal to become U.S. Coast Guardsmen is a clear
indication of how much they cared for their profession. As time passed, the civilian
keepers grew old and retired.

IN RECENT YEARS, there has been an effort to look deeper into who actually
served as keepers of the lights of this country. Some of the results have shown a
diversity hitherto little known. Reflecting the mores of American society until recent
years, most employees of the U.S. Lighthouse Service were white men. There were,
however, some very notable exceptions to this generalization. African Americans
played a role in the service that has been largely unrecorded. The first mention of
an African American in connection with lighthouses appears in a report concerning
a boating accident. Keeper George Worthylake, his wife, and daughter, drowned in
1718, just two years after becoming the first keeper of Boston light station. Suffering
the same fate as Worthylake was his unnamed slave.[51]

An unusual event involving an African American in the U.S. Lighthouse Service
took place early in 1836 during the Seminole War (1835–1842). A raiding party
of Seminoles struck a homestead thirty miles from Cape Florida, killing five settlers.
Military planners felt the next logical point of attack along the southeastern Florida
coast would be at the isolated Cape Florida lighthouse. The sixty-five-foot-high brick
structure guided mariners past the dangerous Florida Reef and into the Cape Florida
channel. Hurricane season was approaching, and it was important to keep the light
shining. Assistant Keeper John W. B. Thompson and one "nameless Negro helper
volunteered to remain at the lonely tower until the keeper could arrange to have
troops sent up from the garrison at Key West."[52]

As expected, Seminole warriors struck the lighthouse on the afternoon of 23 July 1836. When the warriors were only twenty yards away, Thompson later related that he "called out to the old Negro man" that was with him, and both men ran to the tower. Barring the door, the keeper stationed the helper near the entrance while he gathered up three muskets loaded with ball and buckshot. Thomson went to a window and began firing at the milling Seminoles. The warriors took cover near the lighthouse and waited for darkness.[53]

Once the sun set, the warriors struck. Some of the raiding party managed to creep close enough to set fire to the lighthouse's tower doors, while other warriors fired at the tower. Musket balls managed to strike the oil storage tanks, soaking everything, including the two defenders' clothing, in the flammable liquid. Flames began to lick at the oil. Thompson yelled to his assistant to take to the stairs. The keeper

Keeper Malone (so-named by lighthouse commissioner George Putnam) learned that the Isle Royale Light Station in Lake Superior was for married couples only. So he went to the mainland, married, and moved back to the island, where he and his wife raised twelve children.

grabbed weapons and powder and followed his assistant. Both men barricaded the stairwell with the scuttle to the lantern room.

The scuttle held off the creeping flames for a time, but then, according to Thompson, "the crackling flames burst around me. The savages at the same time began their hellish yells. My poor Negro looked at me with tears in his eyes, but he could not speak."[54] Their clothing quickly caught fire, and the two men appeared doomed. They would burn to death, or be cut down by the warriors. Trapped and in extreme pain, Thompson apparently thought to end it quickly. He grabbed a keg of gunpowder and heaved it down the flaming stairwell. The explosion rocked the tower, but it held together. Unexpectedly, it dampened the fire briefly. A musket ball hit the African American and he called out, "I'm wounded," and then died.[55]

Thompson, suffering and alone, still trying to end it all, dragged himself over to the edge of the gallery and prepared to leap from the tower. Just then, the flames died and a breeze sprung up giving the burned and battered keeper some relief. The warriors felt the combination of fire and explosion must have killed both defenders and left the site.

Somehow, Thompson managed to hold on until the next afternoon when he spotted the USS *Motto* lying offshore. The keeper tore a scrap of clothing from his African-American assistant's bloody trousers and waved it to attract attention. The *Motto* landed a detachment of sailors and leathernecks. Thompson's ordeal had not yet finished. With the stairwell destroyed, the keeper had no way to make his way down the sixty-five-foot tower. A musket ramming rod was fixed with a light line and then fired to the top of the light tower; Thompson used this to eventually bring a heavy line to his position, thereby finally ending the long ordeal.[56]

One historian notes that the records of the U.S. Lighthouse Service, reflecting the nature of the times, rarely recorded the presence of African Americans. Other than the dramatic incident at Cape Florida, African Americans are most mentioned in inspection reports or when a rule was violated. In 1852, the U.S. Lighthouse Board noted the New Point Comfort, Virginia, lighthouse was "manned by a retired sea captain and his assistant—a female Negro slave." The keeper of the Cape Hatteras lighthouse lost his position when it was discovered that he was "using his Negro slaves to tend the light." African Americans did serve in some lightships, as an 1835 regulation "specifically forbade their hiring aboard lightships except as cooks." At present, there is no strong evidence an African American ever served as a keeper.[57] A complaint in October 1826 by a Mr. Pinkney, collector of customs at Key West, however, suggests something interesting. Mr. Pinkney wrote that the keeper of the Cape Florida lighthouse "does not live in the dwelling attached to the lighthouse, nor has he for some time past. . . ." The collector then went on to remark that the keeper "built a house on the mainland several miles from the lighthouse and . . . [gave] the whole direction of the light to a black woman."[58]

If the history of African Americans in the U.S. Lighthouse Service is not heavily documented, then evidence of Native Americans working at lighthouses must be considered extremely slight. Only a few mentions exist in correspondence of the U.S. Lighthouse Service. The keeper of the Gay Head, Massachusetts, lighthouse, for example, hired Native Americans instead of African Americans, because "he considered them more competent."[59] Interestingly, there is revealing information in the files of the U.S. Life-Saving Service that hints of the employment of Native Americans at light stations and of their bravery. One example relates to a rescue of the British ship *Lammerlaw* off the present-day Washington coast in 1882.

At 0500 hours on 30 October 1882, the iron bark *Lammerlaw,* laden with 1,125 tons of coal, fought through "very heavy seas" and "frequent squalls," bound for Portland, Oregon. The skipper of the ship mistook the Shoalwater Bay, Washington, light for that at Cape Hancock; the ship struck the south end of the northern breakwater at the entrance to the bay. Two hours later, Albert T. Stream, the keeper of the Shoalwater Bay U.S. Life-Saving Station, went out to a hill to check the condition of the bar with a long glass; he could see nothing because a large squall passed just at that very moment. As Keeper Stream made his way toward the Shoalwater Bay Light Station, Keeper Sidney Smith met him and told him about glimpsing a ship on the rocks.

Shoalwater Bay, in the words of the official report of the U.S. Life-Saving Service, "is quite isolated, there not being more than four or five white men within a radius of ten miles. The main reliance [for crewmen on a boat rescue] was upon an Indian village about two miles from the station. . . ." Keeper Stream sent Keeper Smith to the village to obtain help. Three men volunteered to go out into the heavy weather with the keeper. Smith knew this would not be enough help, so he told the men not to come; he then somehow managed to make a journey of fifteen miles to the nearest settlement and returned to the Shoalwater U.S. Life-Saving Station with a tug and "some men." They arrived back at about four in the afternoon. In the meanwhile, Keeper Stream, along with Assistant Lighthouse Keeper Tilben and "several ladies" from a nearby mission, managed to launch the surfboat and made ready to face the raging sea.

Seven men clambered into the surfboat, four from the crew of the tug. The volunteer crew took a tow line from the tug and put out into the heavy seas on their "stormy journey." Darkness now entered the picture. The skipper of the tug refused to go any further in the "wild night" and demanded that his original surfboat crew come aboard the tug. He cast the boat loose "leaving the brave keeper no course" but to pull against the tide to the station. The three-man boat crew returned at eight that evening. Keeper Stream ordered the men to bed, telling them to be ready "for action" in the morning. The keeper then went out into the night and stood where he could be seen by the sixteen sailors aboard the *Lammerlaw.* Stream then burned a Coston signal (a flarelike device) to let the hapless crew know that they had been spotted and help would be on the way. He then set out through the darkness to the Native American village and asked if the three men were still willing to go out into the storm. The three—whose names are recorded as Light-house George,

Light-house Charley, and Indian Bob—agreed to go and they returned to the station with Stream.

Keeper Stream later roused out his crew of six men (the three from the previous attempt and the three Native Americans)—and decided to push out into the heavy seas even before the tug returned. At three in the morning, the six men and Keeper Stream positioned themselves in the surfboat and took up their oars and pulled down the bay for six miles against the tide. The seas grew rougher, and Stream decided to anchor until daylight.

"When the day broke," the *Lammerlaw* could be seen a mile away, "with her hull well buried in the turbulent floods, and the surf every moment flying in great sheets over her." The bark's crew could be seen dimly toward the stern, some in the mizzen rigging and others atop the cabin. "All around this lamentable sight the sea ran and burst terribly." At the sight of what awaited them, Light-house Charley and Indian Bob declared "that it would be impossible to get nearer the wreck . . . Light-house George behaved nobly and stood by his oar." The two who demurred and the condition of two other crewmen caused Keeper Stream to await the tug.

Around noon, the tug hove to near the boat and Keeper Stream replaced the two Native Americans and the two other crewmen with four sailors from the tug so as to have a largely rested crew to begin the battle. Light-house George and the other five crewmen were now ready to follow Keeper Stream's orders. The men put their backs into the oars.

"There was positively no lee for the boat's approach" to the *Lammerlaw,* the breakers "ran and volleyed around the hull [of the bark] on every side." The "waters literally raged," the wreck "the center of an abattis of flying chutes and cataracts." The surfboat moved slowly toward the side of the *Lammerlaw,* the crew "keeping a terrible grip upon the oars and straining for their hold against the sea." The boat half filled with water. The crew feverously bailed. An oar snapped. Eight sailors from the bark made it into the small surfboat, two of them dragged through the raging waters by a line.

In the meanwhile, another tug appeared just outside the breakers. Keeper Stream made for this tug. He transferred the eight crewmen, including the injured captain, to the tug and then pointed the surfboat into the raging breakers for another battle with the surf. Another oar broke. The surfboat twice filled, with the crew again quickly bailing. The seas knocked Keeper Stream down, but one by one the remaining crew of the *Lammerlaw* came into the small surfboat. Stream once more made for the tug, which took the boat in tow to the station. For his leadership in this rescue, the British government awarded Keeper Albert T. Stream a medal. It is not recorded whether Light-house George and the rest of the brave crew received similar recognition.[60]

Another Native American is mentioned in a rescue attempt near Point Arena, California. On the morning of 22 November 1896, the steamer *San Benito,* laden with four thousand tons of coal, came ashore about two hundred yards from the beach and about four miles north of Point Arena. The surf was running extremely high, driven by a southeasterly gale. The force of the swells and striking bottom caused the ship to break in half, the two sections lying about twenty-five yards apart. Many of the crew took to the rigging, but eight or nine men managed to launch a lifeboat, which immediately capsized, and several of them drowned.[61]

Shortly after this, Jefferson M. Brown, the head keeper of the Point Arena Light Station arrived to see if he could help. Seeing the men in the rigging and knowing that the remaining crewmen were in great danger, Keeper Brown began to try and find volunteers to help him with the capsized ship's boat and attempt to save the sailors. Even though "there were hundreds of people" on the beach, it was difficult to find anyone willing to brave the surf. The onlookers had good reason to demur: the *San Benito* sailors who had survived the capsizing "condemmed the boat saying she was 'no good' and that an attempt to reach the ship with her would be suicidal through such surf as was running. . . ." Keeper Brown, however, kept haranguing the crowd and managed to recruit Lazar Poznanovich and a Native American known as "Sam," later identified as Sam Miller.

The lighthouse keeper and his two volunteers, with help from the onlookers, launched their boat into the raging surf. Keeper Brown manned the sweep. The Brown's boat played out a line from shore, apparently with the notion that if the boat capsized, the crew could be pulled ashore. The stalwart crew managed to make it through some of the breakers, about halfway to the men huddling in the rigging of the *San Benito,* when the current set them north of the wreck. Witnesses stated that "it seemed as though the boat would be swamped" several times during this attempt, and it was "a matter of the greatest doubt" that the rescue crew would ever reach shore again. The rescue boat, after being set north of the wreck, was pulled ashore.

Keeper Brown again called for more volunteers to help, now knowing three men had little chance against the seas and current. Again, no one stepped forward, especially after "everyone had seen how the boat was tumbled by the rollers." Finally, two additional men volunteered to help make another attempt, with a line being let out over the stern as before. This attempt brought the boat to within thirty or forty feet of the wreck, but was again set to the north of the desperate sailors in the rigging. The crowd on the beach again pulled the boat ashore. The rescue boat looked as if it might capsize, and Keeper Brown's sweep oar was broken from the force of the seas.

Brown's third attempt for more volunteers was met by silence from the crowd. The two extra men who had made the second attempt refused to go again. As Brown was trying to gather extra help, a sailor on the wreck, deciding no one could make it through the heavy seas, jumped into the ocean in a desperate attempt to make shore. The unfortunate sailor could not make the distance and drowned "in plain view of hundreds on the beach."

Keeper Brown now knew it was an impossible task to get the heavy boat through the seas with just two men and was forced to give up his attempts. On the afternoon of 22 November, the steamer *Point Arena* hove to near the scene. The steamer's boat crew, after treacherous work, managed to get six sailors from the rigging. The seas were so bad that the remainder of the sailors could not be brought to safety until the following day.

For his heroic efforts, Keeper Jefferson M. Brown was awarded the highest medal of the U.S. Life-Saving Service for rescue at sea, the Gold Life Saving Medal. The service, noting that Keeper Brown could not have accomplished anything by himself, also award the Gold Life Saving Medal to Lazar Poznanovich and the Native American Sam Miller.[62]

All maritime professions in the United States until well into the twentieth century were the providence of men. The U.S. Lighthouse Service, however, did employ a surprisingly large number of women at lights. Mary Louise Clifford and J. Candace Clifford, in their book, *Women Who Kept The Lights,* note at least 240 names of women who were assistant keepers and 138 appointments of women as head keepers. However, this employment of women did not spring from an enlightened equal employment policy, far from it. The service merely took advantage of the labor available at lights that was provided by families. In the great amount of spare time at isolated stations, women and children could learn the business of tending a light. At times, it became almost a necessity. Often, the father or husband had to service minor lights, and someone had to tend the main light if he became delayed. Too, the head keeper needed to leave the light from time to time for supplies and, again, a family member could tend the light in his stead. In these cases, the wife or children acting as assistant keepers received no pay. In other cases, the wife would receive a salary, but at a lower rate than the husband. The service, in short, could get two or more employees either for the price of one or one at a reduced wage.[63]

A death of a husband or father could lead a woman to become the head keeper. Frequently, this lasted only a few months until a replacement keeper arrived, and then the wife or daughter had to leave the station. The service sometimes made the tenure permanent, as there was no pension or compensation for wives or daughters. Sometimes no man thought that particular station a good assignment, and a wife or daughter would then take over the permanent position.

Sketches of only a few of the women employed by the U.S. Lighthouse Service follow, because of the numbers involved. Most women keepers, like men keepers, were people of average abilities who sought no special mention. They did their work with very little, if any, notoriety. A few, however, deserve special mention because they accomplished something out of the ordinary for their times. The brief introductions to these women illustrate what set them apart from the average keeper.

Abbie Burgess was one of the more articulate lighthouse keepers. Samuel Burgess received the appointment of head keeper of the Matinicus Rock Light Station in 1853. The station stood on a rock outcropping, four miles from the south end of Matinicus Island, Maine, and twenty miles from the mainland. Samuel brought with him his invalid wife and five children. The son, Benji, spent most of his time away from the island fishing. Abbie, born in 1839, was the eldest of the daughters.[64]

The station had two stone towers, a structure for a fog bell, and other smaller buildings for storage of oil, boats, and other equipment. A cistern collected water. The keeper's rubble-stone living quarters connected the two towers.[65]

Abbie learned to tend the station's twenty-eight Argand lamps. In January 1856, supplies at the station were running low, so Keeper Burgess left the island to buy food. With Benji pursuing his fishing career, Abbie now had charge of the station. A storm struck Matinicus Rock, stranding Samuel ashore. Meanwhile, on the island sheets of spray, whipped up from the angry seas, flew over the rock. Snow and sleet lashed the lighthouse. Abbie moved her invalid mother and three sisters into one of the light towers. At high tide, the seas breached the island. In a letter to a friend, Abbie later wrote the "dwelling was flooded and the windows [shutters] had to be secured to prevent the violence of the spray from breaking them in. As the tide came in, the sea rose higher and higher, till the only endurable places were the light towers. If they stood we were saved, otherwise our fate was only too certain. . . ."[66]

The weather proved so stormy that no boat could land at the island for a month, but, as Abbie wrote, "not once did the lights fail."[67] A wave that swept the rock and destroyed a building almost struck the young assistant keeper as she tried to rescue the family's chickens. A year later, Abbie again singlehandedly tended the lights, when another storm kept her father from the rock for three weeks. This time, the food supplies were down to a meager one egg and one cup of corn meal a day.[68]

Abbie's father lost his position to a Republican appointee in 1860. The young woman remained on the island to operate the light until the new keeper learned his duties and ended by marrying the replacement keeper's son, Isaac H. Grant. Abbie then became the new keeper's assistant, receiving a salary of $440 a year. Abbie and her husband remained on Matinicus Rock until 1872. The couple then transferred to White Head Light Station, Maine, where again Isaac was head keeper and Abbie assistant keeper; her salary eventually reached $480 annually. Isaac Grant died in 1875, and Abbie received the appointment of principal keeper in that year. Abbie Burgess Grant continued as keeper in 1889, "and may have kept her post until she died in 1892," spending thirty-eight years of her life, and raising four children, at a lighthouse.[69] One of her daughters, Bessie, died on Matinicus Rock. The best that Abbie and her husband could do for a grave was "to scrape together a few handfuls of soil, make a wooden cross and bury" her between rock ledges. Almost a hundred years later, Harriet and Carl Buchheister hired a stone mason to carve a headstone, which the U.S. Coast Guard erected over the baby's grave.[70]

Abbie never seemed to regret her life of isolation. She wrote descriptive and sometimes moving letters of life at a light station. Her last letter before her death eloquently relates her feelings about her life's work: "Sometimes I think the time is not far distance when I shall climb these lighthouse stairs no more. It has always seemed to me that the light was part of myself . . . I wonder if the care of the lighthouse will

One of the stories of Abbie Burgess's time at the light station at Matinicus Island, Maine, concerns her rushing out into a gale to save the family chickens. The effort nearly cost her her life. Charles Nordhoff depicted the moment in his etching titled "Abby Saves the Chickens."

follow my soul after it has left this worn out body! If I ever have a gravestone, I would like it in the form of a beacon."[71]

Edward Rowe Snow, author of many lighthouse books, read Abbie's letters and wanted to view her grave. Snow found no one had honored the faithful lighthouse keeper's wishes concerning her grave stone. Fifty-seven years after Abbie Burgess Grant's death, Snow was instrumental in placing a stone in the shape of a lighthouse on Abbie's final resting spot.[72]

Idawalley Zorada Lewis, better known as Ida, came to the Lime Rock Light Station, Newport, Rhode Island, in 1857 as the fifteen-year-old daughter of the keeper. Hosea Lewis, Ida's father, died in 1872, and his wife became the head keeper, with Ida assisting. Ida married a Capt. William Wilson of Black Rock, Connecticut, in 1870, but they separated two years later. Ida became the head keeper in 1879, after her mother gave up the position. She remained at Lime Rock until her death, forty-six years later, in 1911. Arguably, she is the most famous woman keeper in the U.S. Lighthouse Service. Ida's fame began in her own time. Her exploits appeared in many publications, and Ida was featured on the front cover of the 31 July 1869 issue of *Harper's Weekly*.[73]

The daughter of Captain Hosea Lewis, a coast pilot, Ida entered life in 1842 at Newport, Rhode Island. Captain Lewis, suffering declining health, "swallowed the anchor," and became the first keeper of the new Lime Rock Light Station in 1853. Hosea lived in a temporary crude shelter until 1857, when he moved his family into the new keeper's quarters.[74]

Captain Lewis suffered a crippling stroke. Ida and her mother began tending the light in his stead. Ida also helped care for her father and a seriously ill sister. The eldest of the four Lewis children, Ida had the additional duty of transporting her younger family members to and from school and bringing supplies to the island. These duties enabled Ida to gain a great deal of skill at handling the oars of a heavy boat. In an age where such a skill drew comments of "unfeminine," Ida could, and did, handle a pair of oars better than many men. Ida constantly proved her mastery of an oar-powered boat. Ida's brother bragged that she could "hold a boat to wind'ard in a gale better than any man I ever saw wet an oar. Yes, and do it, too, when the sea is breaking over her." During her first year on the rock, Ida rescued four young boys after their boat capsized. Over the next thirty-nine years at Lime Rock, the official record credits her with saving eighteen lives, although some unofficial sources suggest at least twenty-five rescues. On 4 February 1881, two soldiers from nearby Fort Adams were crossing the ice at around five in the afternoon, when they hit a weak spot and plunged into the ice-choked water. Ida heard their cries and, not hesitating for a moment, grabbed a rope and rushed out onto the ice, even though the "ice was in a very dangerous condition." Nearing the spot and "in imminent danger of the soft and brittle ice giving way beneath her," Ida threw the rope to the soldiers. When hauling in the first man, she faced the danger "of being dragged into the hole by the men," because both panic-stricken soldiers grabbed

Ida Lewis was arguably the most famous woman lighthouse keeper. She lived at the Lime Rock Light Station at Newport, Rhode Island, from 1857 until her death in 1911. She became head keeper in 1879. Ida, noted for her ability to handle a boat and for her many rescues, was awarded the Gold Life Saving Medal in 1881.

the line at once. Somehow Ida managed to persuade one man to let go, and she hauled the other to safety. By this time, her brother arrived and helped in pulling the other sodden soldier from the water. The U.S. Life-Saving Service noted Ida's action "showed unquestionable nerve, presence of mind, and dashing courage." This act of heroism caused the Treasury Department to award her a Gold Life Saving Medal. At the age of sixty-three, Ida performed her last recorded rescue. When asked how she managed to find the strength, Ida replied, "I don't know, I ain't particularly strong. The Lord Almighty gives it to me when I need it, that's all."[75]

Ida's fame spread throughout the country. Many people made the journey to the light to see the prominent keeper. Her wheelchair-bound father counted the number of people visiting the island; Hosea recorded nine thousand visitors one summer.

The crowds made Ida uncomfortable. She used Newport journalist George D. Brewerton to answer her most frequently asked questions, when he wrote a pamphlet about Lime Rock. President of the United States Ulysses S. Grant and Vice President Schuyler Colfax both visited with Ida. Her fame also brought some monetary reward. The Life Saving Benevolent Association of New York awarded her a silver medal and a check for $100. Newport held a parade in her honor, followed by the presentation of a mahogany rowboat. At the age of sixty-four, Ida became a life beneficiary of the Carnegie Hero fund, which carried a monthly pension of $30.[76]

Once asked how she managed the difficult duty of watching the light every night, Ida replied, "The light is my child, and I know when it needs me, even if I sleep."[77] Ida remained at the Lime Rock Light until her death in 1905. Her gravestone in the Common Ground cemetery has crossed oars etched upon it.[78] In 1924, the Rhode Island legislature changed the name of Lime Rock to Ida Lewis Rock. The U.S. Lighthouse Service changed the name of the light station to Ida Lewis Light Station, the only light named after a keeper.[79] The U.S. Coast Guard paid her the honor of naming the first of a new class of buoy tender the *Ida Lewis*.[80]

To the south of Newport, in the shadow of New York City, Kate Walker served from 1883 to 1919 at a confined caisson light station. Kate Walker's association with lighthouses actually began at the Sandy Hook Light Station in New Jersey where John Walker worked as an assistant keeper. He took his meals in a nearby boardinghouse. Kate, a young immigrant woman from Germany with a small boy, waited tables at the house. John decided to teach her English; he ended up marrying her as well. John took his new bride to Sandy Hook and taught her how to tend a light.

John received the head keeper's appointment to Robbins Reef Light Station in 1883. Kate accompanied him and received the assistant keeper position. She later related that, upon viewing the narrow confines of her new home, she threatened to leave John. "I refused to unpack my trunks at first, but gradually, a little at a time, I unpacked." In a marvelous piece of understatement, she added, "After a while they were all unpacked and I stayed on." She indeed did stay on—for thirty-six years.[81]

Robbins Reef Light Station marks a reef on the west of the main channel into New York's inner harbor and about a mile from Staten Island. The reef upon which the lighthouse sits is below water, so the service constructed a masonry island within a caisson. The main level of the light, which held part of the keeper's quarters, fit around the base of the fifty-six-foot iron tower "like a donut." The light had no area for a boat house; the station boat, hoisted by davits, hung at the main level of the light—the "donut." Access to the light from the water was by a vertical ladder up to the kitchen door. The main level contained the kitchen and dining area. Clothing and china lockers were on the first floor of the iron tower. Two bedrooms were on the next level.[82]

Three years after taking over Robbins Reef Light Station, John was struck down with pneumonia; he did not survive the illness. Most accounts of the lighthouse's history relate that his last words, before being taken by boat away from the light,

From 1883 on, the light station at Robbins Reef was the home of Kate Walker. She was made its head keeper in 1886 and remained so until her retirement in 1919. Legend says she saved the lives of at least fifty people.

were, "Mind the light, Kate." Kate managed to obtain a substitute so she could attend her husband's funeral, but she returned to her light by that evening.[83]

The service offered the position to several men, but they refused saying the light was too lonely. Kate, now forty years old, quite logically applied for the position. The service hesitated. Apparently, decision makers thought that the four-foot, ten-inch tall widow, who weighed barely one hundred pounds, could not do the job. Nevertheless, Kate eventually became the head keeper of the light station, with her son Jacob as assistant keeper. Her size never hindered her abilities to tend the light or to help others in distress. According to her own account, she rescued at least fifty people, mostly fishermen blown onto the reef by storms.

In winter, one of her most pressing duties entailed removing frost and snow from the glass windows of the lantern room. Kate braved the elements to work outside on the balcony to clear the snow from the windows. When visibility dropped, she started the fog signal, whose blasts prevented any chance of sleep. If the fog signal machinery broke, Kate would go to the balcony to ring a fog bell. If the nearby Staten Island Depot heard the bell ring out, they knew to send a repair force to Robbins Reef.

Kate Walker retired in 1919. She lived for another twelve years on Staten Island, dying at the age of eighty-three. A New York newspaper's obituary aptly described the woman who kept a light that many men felt too lonely: "amid . . . sight of the city of towers and the torch of liberty lived this sturdy little woman, proud of her work and content in it, keeping her lamp alight and her windows clean, so that New York might be safe for ships that pass in the night."[84]

Across the continent, Laura Hecox loved the marine life she found while accompanying her father along the northern edge of Monterey Bay, California. Laura was born in 1854 in Santa Cruz, California. As a young woman, she explored the tide pools, gathering samples of rocks and minerals and collecting sea shells. Laura's father, Adna, became keeper of the Santa Cruz Light Station in 1869. This allowed Laura to live close to the shoreline she loved to explore. The young woman came from a large family: she was the ninth of ten children. Following the manner of many keepers, Adna taught his children how to help in tending the light.[85] (He taught only the two youngest as the others were grown; Laura was one of them.)

The life at the Santa Cruz light allowed Laura to become an ardent shell collector. By the time she reached her twenties, Laura was in contact with collectors and exchanging specimens. Adna's health began to fail, and Laura took over more of the duties of maintaining the light.

Laura's father died in 1883. Her brother-in-law wrote the U.S. Lighthouse Service that Laura knew the duties of tending the Santa Cruz light and logically should be the next keeper. Within a week, the appointment became official and, at the age of twenty-nine, Laura became the head keeper of the Santa Cruz Light Station. The lighthouse for the first forty-eight years of its existence was really the Hecox Family Light Station. Three of Laura's family were married at the light and three died there. One of Laura's brothers came back to live. A sister and her husband also returned. Laura's mother lived at the light until she died in 1908, at the age of ninety-two.

Lighthouse duty allowed Laura to continue with her avocation of collecting natural history specimens. Over the years, the light station became a magnet for visitors wanting to see both the workings of the light and the large collections of shells and other marine artifacts. In 1904, only Laura and her mother lived in the six-room keeper's dwelling. Laura turned one room into a private museum for the natural history collections. The room also contained historical artifacts and a library of scrapbooks on wide-ranging topics. When a new library opened in Santa Cruz in 1902, Laura donated the entire collection for permanent display. The Hecox Museum opened in 1905.

Laura Hecox lived at the Santa Cruz Light Station for forty-eight years. She probably worked the light for all those years, beginning by first assisting her father.
Officially, she worked as the head keeper for thirty-four years. Laura retired in 1917
at the age of sixty-three. She died two years later.

Laura Hecox's tenure at the Santa Cruz Light Station is without controversy. She
performed her duties as a lighthouse keeper in the traditional and praiseworthy manner of all keepers. In addition, this woman who apparently had no chance for higher
education, found and perfected a strong interest in marine biology. Her interests
helped a museum provide interesting artifacts for others to enjoy.

FOR MANY YEARS, the U.S. Coast Guard planned to automate all the lighthouses
in the United States. In 1989, however, Congress decreed that Boston light, the
oldest in the United States, should "remain permanently manned."[86] For all intents,
however, a part of maritime history has passed, little noted by most Americans. Since
time out of mind, people have tended lighthouses so that others could safely reach
their destination. In the United States, lighthouse keepers, at first, were not the best.
Under the direction of the U.S. Lighthouse Board, however, America's keepers
became a professional group that ranked among the best, if not the best, in the
world. Keepers and their families lived a life of isolation and monotony that is difficult for modern Americans to comprehend. Because of the nature of their work,
they also faced the chance of death. They did their work day in and day out with
very little, if any, notice from their fellow citizens. In recent years it has become popular to romanticize keepers and, indeed, it is easy enough to follow this path. It is
important, however, to realize that lighthouse keepers were ordinary men and
women who stayed by their lights and did the best job they could do under trying
circumstances. Some performed their duties nobly, others had feet of clay, but anyone whose ancestors came to this country by sea, or who went down to the sea in
ships, and reached port safely owes these ordinary men and women a great debt.

Ghosts and the Places They Haunt

ONE AUTHOR HAS CLAIMED that hardly "a lighthouse exists that does not have some supernatural being associated with it."[1] The entire milieu of lighthouses seems ready made for ghosts and strange happenings. Isolated locations, storm-tossed waves, wrecks and loss of life, beams of light-piercing, fog-blanketed darkness, the mournful bleat of a fog horn, all seem scripted by Stephen King. Add to this the lively imagination of keepers who passed the long, isolated hours telling stories, then mix with locations noted for some macabre tales, and it is not difficult to understand why the stories of ghosts and strange happenings became part and parcel of the history of the U.S. Lighthouse Service. Some tales probably were in the same spirit as sending a "boot" sailor looking for green oil for the starboard running light, or looking for a left-handed monkey wrench. The stories do illuminate a part of the lives of those that lived with the lights, although one that cannot be documented in the traditional manner of historians. What follows is a sampler of stories that have managed to work their way into the folklore of the U.S. Lighthouse Service and a few that have become entrenched in U.S. Coast Guard history.

Probably just as important as ghost stories were those that no doubt began with some preface, such as "Did you hear the story about . . . ?" Take, for example, the story of a strange Thanksgiving. The keeper prided himself in preparing a large holiday dinner. This particular Thanksgiving eve the weather did not cooperate. Storms kept the lighthouse tender from arriving, and the larder was at a bare minimum. That night, the head keeper heard a loud thump in the vicinity of the lantern room. He rushed up the tower and looked out the glass in the lantern room. On the balcony lay four pairs of black ducks, dead from striking the lighthouse. Glancing down from the tower, the grateful keeper spotted four more ducks lying on the ground. Thanksgiving dinner was saved.[2]

Pets were one way keepers could help ease the loneliness of their stations, and stories of animals circulated throughout the service. Spot, a dog at Two Bush light in Maine, helped saved the lives of two fishermen. A strong nor'easter caught a fishing boat with two men aboard before they could reach shelter. The fishing craft rammed into a ledge at eleven at night, and the two fishermen took to their dory. For the next three hours, the men fought the raging seas. Ice and sea water began to make the small boat loose its buoyancy.

Then, over the sound of the wind and seas, they heard the sound of a barking dog. "It sounded to us like the voice of an angel," said the skipper of the fishing vessel.

Spot's barking alerted the keeper of the light that something might be wrong. The keeper went to where Spot was running up and down the beach and there he spied the two men shouting for help. Eventually, the keeper guided the two men to safety. The rescued men tried to buy the dog, but the keeper refused.[3]

Yet another dog by the name of Spot resided at the Owls Head, Maine, light. This Spot, a springer spaniel, loved the station's fog bell. The dog learned how to pull the bell's rope from watching the keeper. Spot's favorite game was spotting ships and dashing over to the bell and pulling the rope apparently to salute the passing ship. Spot would ring the bell until the ship would sound its horn and then run down to the water's edge to bark until the ship was out of sight.

One stormy night, Mrs. Stuart Ames, wife of the skipper of the mail boat running between Rockland and Matinicus, called the light station to ask whether the keeper had spotted the mail boat, as it was several hours overdue. The keeper told Mrs. Ames that a strong snow storm prevented him from even seeing the water. Desperate, Mrs. Ames told the keeper that her husband said Spot never missed a chance to welcome the passing mail boat. Would the keeper mind letting Spot outside into the storm to listen? The keeper agreed and out Spot rushed, only to return, wet and cold. The springer spaniel entered the keeper's quarters and curled up by the fire. Suddenly, the dog sprang to its feet, rushed to the door and began scratching to be let out. The dog ran to the fog bell, but, because the snow was so deep, Spot could not find the rope. He then ran to the beach and began barking frantically. Out of the gloom came a blast from the mail boat. Spot's barking became even more frantic. Then three, sharp blasts let Spot and the keeper know Captain Ames had heard the dog and knew the safe channel to port. Two hours later, Mrs. Ames called to say her husband had made it safely home, thanks to Spot. According to one chronicler of Maine's lighthouses, Spot was buried near the fog bell he loved.[4]

Of course, these stories were tame, and probably a keeper would warm up his audience for ghostly deeds by telling of an incident that happened near Mount Desert Rock, Maine. A keeper watched a fisherman hauling in his nets. Next morning, the keeper was amazed to see the boat still in the same spot. Thinking something might be wrong, the keeper launched the station's boat and rowed out to the trawler. He went aboard the craft and found no one there. Then he noticed a line leading into the water and decided to haul the line into the boat. The keeper began to pull in the line and found a halibut weighing well more than a hundred pounds. He had another grisly catch as well: the fisherman's body. Apparently when the large halibut hit the line, the fisherman was hauling in at the same time. The fish pulled the line out, and a hook sank into the fisherman's hand dragging him to his death.[5]

Even some *locations* of lighthouses had lurid stories attached to them that would occupy the minds of keepers. At Boon Island, Maine, for example, the *Nottingham Galley* wrecked there in 1710. Before help could arrive, the crew resorted to cannibalism to survive.[6] There is the story of the pirate Edward Teach, better known as Blackbeard. Teach used Ocracoke, North Carolina, as a haven because it was an extremely difficult area for a ship to enter. An English naval lieutenant named Maynard, however, managed the feat and surprised the pirate. Maynard dueled with Blackbeard and won. He decapitated the pirate and placed the bloody head in the bowsprit. Supposedly, Blackbeard's body swam three times around Maynard's ship

Reportedly, colonial prisoners were condemned to death by drowning from the site of the Execution Rocks Light Station.

before sinking. Keepers of the Ocracoke Light Station were among those who passed on the legend of Teach's headless ghost swimming around the cove looking for its head and then coming ashore with a lamp.[7]

Another unusual story concerning the location of a lighthouse is that from Execution Rocks. The lighthouse at Execution Rocks warns mariners who enter New York's harbor by way of the East River from Long Island. One may pick either of two reasons for how the rocks received their dreadful name, both of which fit nicely into U.S. Lighthouse Service folklore. The two are probably variations of one theme. One version claims that during colonial times when prisoners received the death penalty, judges would order the prisoners to be taken out to the rocks and chained to a ring bolted in a pit at the low-water mark. The prisoner was left there to await his fate as high tide inundated him. The second version has it that, during the American Revolution, the British would chain American revolutionaries (who at public executions would shout inflammatory remarks and stir the gathered crowds) to the rock below high tide to await drowning with the incoming tide. The ghosts of these revolutionaries supposedly reaped their revenge. A boatload of British soldiers, in hot pursuit of Washington's retreating army, for some unknown reason veered suddenly toward Execution Rocks, "as if pulled by unseen spirits"; they wrecked on the rocks and all aboard drowned.[8] According to one source, when the light at Execution Rocks went into operation in 1867, a keeper could ask for transfer from the station any time. "Never again, ruled Congress, would any man feel 'chained' to Execution Rocks."[9]

At Crescent City, California, one keeper claimed the lighthouse became the site for a visitation of "miniature mermaids." The keeper, perhaps remembering the old sailor's song, "Eddystone Light," befriended the wee mermaids who, according to the keeper, eventually came into the lighthouse.[10] For those not familiar with the song, a few lines will help.

> *My father was the keeper of the Eddystone light,*
> *And he slept with a mermaid one fine night.*
> *From this union there came three,*
> *A porpoise, a porgy, and the other was me. . . .*

At Sandy Hook, New Jersey, a new keeper to the light began his duties sometime in the nineteenth century. Upon examining his new living quarters, the employee noticed the building seemed to have a basement, but he could find no entrance to this sublevel room. As the days passed, this fact worked more and more on the man's mind. Finally, curiosity got the better of him, and he managed to pry up the floor boards and found an entryway into the basement. The keeper warily descended into the lower region. There to his horror, propped up at a table facing a crude fireplace, sat a skeleton. According to folklore, the "mystery of the skeleton was never solved."[11]

The tragic end of the two assistant keepers when the first Minot's Ledge Light Station collapsed in a severe northeast gale brought forth a number of stories, providing a good introduction to ghosts in lighthouses. In the new stone tower, the head keeper kept watch at night at his station in the watchroom. The story relates that his mind naturally drifted to the collapse of the original tower. He leaned over and tapped his pipe against a table. Then, surprisingly, he heard an answering tap, almost like code, from below. The keeper tapped again and received yet another answering response.

Thinking his assistant made the rap to show that he was awake and making preparations to relieve him, the head keeper sat back and waited for the assistant to arrive. Time passed, but the assistant keeper did not appear. Finally, the keeper rang a bell, which was the normal way to arouse the relief and, shortly after that, the assistant arrived. The head keeper naturally asked why he had to ring the bell, especially after the assistant had already tapped that he was awake and preparing to relieve the watch. The assistant replied that he had only awoken after he heard the bell. Both men became alarmed when they realized that the "tapping from below had been the signal customarily used in the old tower where the two men perished in the gale of [18]51."[12]

Keepers of the Minot's Ledge Light Station were not the only ones who reported strange occurrences at the station. Sailors passing nearby the light began insisting that they heard mysterious noises, and some even saw figures clinging to the lower section of the ladder leading to the door of the light. One sighting reported a man clinging to the lower rungs of the lighthouse ladder, dripping wet, waving his arm, and shouting in a foreign tongue. Portuguese fishermen passing Minot's Ledge understood the language; they claimed the apparition shouted, "Keep away, keep away!" The ghost of the young assistant keeper Antoine (who was Portuguese) was usually seen and heard to warn his countrymen just before northeast storms.[13]

Some of the ghosts of Minot's Ledge seemed helpful. The head keeper one morning noted that the lens was brightly polished, even though the assistant keeper, who normally undertook this duty, was still asleep. When the head keeper commented to the assistant about the good job, the other man said he had not gotten around to doing it yet. The following week, the assistant keeper trudged up the tower to clean the lens and again it was found sparkling clean.[14]

Close to Minot's Ledge, Boston Light Station is the home of the "Lady in Black," allegedly the shade of a beautiful southern belle who, during the Civil War, learned her Confederate officer husband was a captive on nearby George's Island, not far from the light station. The woman set out to rescue her husband. The mission went awry, however, and she died before a firing squad for spying. For some reason, which is not clear, her spirit haunted the lighthouse keepers.[15]

The New London Ledge Light Station in Connecticut is inhabited by the ghost of a jilted husband. In 1936, a keeper known only as Ernie found that his wife had run off with a Block Island ferry boat captain. Ernie apparently could not bear this betrayal. He made his way to the lantern room, opened the steel door to the lantern gallery, and leaped to his death.

Keepers of the light thereafter claimed that the steel door would mysteriously open. One U.S. Coast Guardsman related that the "wind can be blamed for many of the strange noises we hear, but I know the sound of that door opening up and no wind ever did that." Others have heard the decks of the light being swabbed, while others have heard footsteps of unknown origins and have felt cold spots in different areas of the structure. Apparently Ernie also has a mischievous side. The fog signal and light would occasionally be switched on unexpectedly.[16]

Another Connecticut light has a bizarre story. On Christmas Eve 1916, Fred Jordan set out in a small boat for his time away from the Penfield Reef Light Station. Jordan's boat capsized in a raging storm, and the sea claimed another victim. When the body of the unfortunate keeper was recovered, a note was found in Jordan's pocket, apparently a reminder to himself to make an entry in the station's log. Later keepers related seeing the figure of a man, dressed in white, moving out of a room in the light down the stairs and then going outside. When the keepers investigated the room the wraith had visited, one of the station's old logbooks was found open to the page dated 24 December 1916.

Even stranger, on some stormy nights, a spirit was seen on the rail of the gallery or flitting about the base of the rocks. A power boat in trouble was supposedly piloted to safety by a strange man in a rowboat who mysteriously disappeared. Two boys were dragged by a man from the water after their boat capsized and were deposited at the base of the light. When the boys went into the lighthouse to thank the man, no one was to be found.[17]

At the St. Simons Light Station in Georgia, Keeper Carl Svendsen and his wife kept a dog named Jinx. Every evening, when she heard him making his way down the stairs, Svendsen's wife would put the keeper's dinner on the table. One evening Mrs. Svendsen heard her husband coming down the stairs and began the evening ritual. She noticed, however, that Jinx for some reason was barking and whimpering. When Carl did not appear for dinner, his wife became concerned and went to the stairs only to find that Carl was still in the lantern room. Carl remained skeptical of his wife's account, but then the keeper also heard the sound of the strange footsteps echoing in the tower. Apparently, the Svendsen's learned to live with the sound of mysterious footfalls, but Jinx remained afraid and would hide whenever the curious noise echoed throughout the lighthouse.[18]

At California's Point Vicente Light Station, a haunting, according to many, was solved. From the time the light was first lit in 1926, keepers reported an apparition of a woman who walked about the exterior in a flowing gown. Being a new structure, no ghosts of former keepers or their families should have been in the light, so the premise was the wraith paced the area looking for a lover lost at sea.

A young assistant keeper reported to Point Vincente. Apparently this keeper did not believe in ghosts and began to search for an explanation of the visitations. After a long study, he came up with a solution. The light itself caused what appeared to be a ghost. According to the keeper, as the light rotated it threw out an arc in "a reversed perenthesis causing the ghostly image to appear." Seen from eighty to two hundred yards distance, this reflection looked remarkably like a woman in a flowing gown. Thus, the young keeper felt satisfied that he had used reasoning to explain the haunting. Not everyone, however, believed this solution, and the story of the apparition continued.[19]

At Crescent City, California, the sound of footsteps climbing the tower could be heard, but only during stormy weather. "Interestingly, it was during . . . storms that early light keepers felt their greatest responsibilities and had to repeatedly climb the tower to maintain a close watch over the light."[20]

The first light station in Alaska was in the cupola of the Russian Baranoff Castle in Sitka. Having the honor of being the first light station, it is only natural that the castle should have a ghost. Supposedly a beautiful Russian, daughter of one of the governors, was compelled by her "cruel parents" to marry a man she did not love. On the evening of her marriage the princess disappeared; her friends became concerned and began a search. As might be expected, she was found dead in "her boudoir."

Soon thereafter a spirit began to move in the drawing room and paced the governor's cabinet. Whenever the spirit passed it left "behind a slight perfume of roses." A guard from the U.S.S. *Pinta* was given the job of patrolling the castle to prevent the theft of flags before an upcoming ball. The guard heard footsteps, but saw only empty rooms. Apparently, the legend of the ghost was so strong that the local populace was worried when a U.S. government party, led by former Secretary of State William H. Seward, visited Sitka. The specter, however, did not bother the party, according to the local newspaper, mainly because "the tale had been carefully withheld from them until they were about to depart."[21]

Murders and unusual deaths are an important part of the tales of the old U.S. Lighthouse Service. One of the first keepers at the Sequin Light Station, located on an island at the mouth of Maine's Kennebec River, brought his bride to his station. The keeper's consort was a frail woman who passionately loved music, but proved to be one of those not suited for the lonely, isolated life demanded of the family of a U.S. Lighthouse Service employee. The continuous blanket of fog that encompassed the area quickly depressed her. On the rare days when the murk would lift, the woman would stare wistfully for hours across the water at the lights of the nearby town of Bath.

Seeking a way to ease his wife's utter desolation, the keeper went ashore to purchase a piano. With a great deal of effort, and with help from some Bath residents, the keeper managed to get the heavy piano back to the living quarters of the Sequin Light Station. Ecstatic to again hear music, the woman went to her piano and began to play, only to discover that she had only one sheet of music. The woman apparently was not able to improvise, so she played the same song over and over and over at all hours of the day and night. As soon as possible, the keeper brought additional sheet music to his wife, but she continued to play and play and play the same song almost without stopping, or so it must have seemed to the keeper. He soon began to awaken to the fact that his wife might be losing contact with reality, for she began to lose weight and would not stop the incessant playing of the one song.

The distance from Sequin Island to the mainland was not great. Residents who lived near the water, when the wind was right, could hear the faint notes of a song continually played on the piano drifting over the water. One day the music stopped abruptly—the keeper lost his mind, strangled his wife, and took an axe to the piano. Residents stated that after the murder they could occasionally hear the notes of a piano seemingly drifting in from Sequin Island.[22]

New England and the East Coast do not have a monopoly on strange events in U.S. Lighthouse Service lore. Yaquina Bay Light Station, the light that guided sailors into the harbor at Newport, Oregon, is the site for the strange tale of Muriel Travennard. The lighthouse was in commission for only three years and replaced by another structure some distance away.[23]

The isolated light station on Destruction Island, Washington, is supposedly the haunt of a mysterious animal.

Muriel, born in the late nineteenth century, was left motherless when very young. Her father, a sea captain, often took his daughter on his coastwide voyages. When Muriel reached her teens, the father did not think a life on a ship, exposed to some of the language and actions of the forecastle, was a proper environment for a young woman.

At just about this time, Muriel's father signed on a new crew for a voyage to Coos Bay, Oregon. Muriel was left in the care of a friend in Newport. Her father departed, telling his daughter the voyage should take only a few weeks.

While Muriel enjoyed her new surroundings, the weeks stretched into months. The young woman began to fear that her father had met with some terrible fate. One day, a group of youths, hoping to take Muriel's mind off her missing father, invited the girl to explore the abandoned Yaquina Bay lighthouse. Muriel accepted the invitation.

The lighthouse proved a shambles. The young adults found a strange iron plate in the floor, which gave way to a compartment with a hole dug in its floor. This strange arrangement held the young people for a short period, but then they moved on to explore the rest of the light structure, leaving the iron door ajar. By late afternoon, everyone decided they had had enough of the lighthouse and decided to return home. In the lowering twilight, just as the group started away from the abandoned Yaquina Bay lighthouse, Muriel stopped the exploring party and said that she had left a scarf inside. The young people waited while Muriel dashed inside the lighthouse to retrieve the forgotten item; it should have taken only a minute to do so.

The group of teenagers waited and waited. As time passed, they began to become nervous and started shouting out Muriel's name, with no response. A few of the young people decided to go inside and find her. A quick search proved fruitless, but then two discoveries sent the youths running in terror from the abandoned lighthouse. At the bottom of the stairs leading up into the tower was a pool of blood and a trail of blood droplets that led to the iron door, which had mysteriously closed. The young adults tried the door without success. Now, thoroughly terrified, the teenagers ran home to report the terrible happenings.

A later search could find no trace of Muriel Travennard. The iron door could not be opened. Even efforts with a strong crowbar could not budge the door. No trace of Muriel Travennard was ever found. A dark stain still "marks the spot where her blood was found." Reports still circulate that her ghost can be seen "peering out from a dark [lantern room and] walking the shadowy path behind the lighthouse."[24]

Those who love lighthouses and believe everything about the old aids to navigation will be interested to know that strange things have not ceased even at the automated lights. The U.S. Coast Guard's aids-to-navigation teams still visit active automated lighthouses to maintain or update equipment. The author learned the story of Mary Margret when discussing the two isolated lighthouses lying off Washington State's coast. An aids-to-navigation team works on Destruction Island

and Tatoosh Island (Cape Flattery); the boatswain's mate first class in charge of the team related some unusual events at the two locations.

The boatswain's mate said that the team is careful not to say anything bad about Tatoosh Island, or "weird things" begin to happen. Even in "normal" times, team members can place their tools in one location and when they try to retrieve them, the tools are missing only to be found in another spot. The team has dubbed the ghost "Mary Margret," and she may be the spirit of Mary Margret Mize, who died on 10 August 1908 on the island. She lies buried inside a picket fence near the grave of Francis S. Tisdale, who died on 11 July 1917. The aids-to-navigation team maintains the two graves. The boatswain's mate said that after working on the island he is now "a believer."[25]

According to people who work on the now-automated light, strange things happen at the Cape Flattery Light Station on Tatoosh Island if anyone criticizes the place. (Mary Margret Mize is buried on the right side of the island.)

Stranger still is Destruction Island. This former, isolated first-order lighthouse is not haunted; instead, the island has a mysterious animal. The location is now over-populated by rabbits first brought to the island by keepers, but there are also tales from the few people who have been able to visit the island of sightings of a large animal that moves very fast through the brush.

ARE ALL THE GHOST STORIES and strange happenings true? It is more important to realize the tales do give us a glimpse into some of the stories that passed among a profession that has slipped behind the veil of history.

Lighthouses Go to Sea

THE U.S. LIGHTHOUSE SERVICE had a sizable fleet of ships. The inventory consisted of lightships and tenders. Lightships were an important element in aids to navigation. These special vessels moored in shallow water, or near shoals, where, because of a difficult location, technology could not build a light structure. Being ships, they could move to meet any changing circumstance. The large expense needed to maintain these special craft was the reason the service early on decided that "light-ships should, as rapidly as possible, be replaced by light-houses."[1]

The history of lightships stretches from the eighteenth century, although there are indications that the pre-Christian Romans employed the first lightships when they placed fire baskets in galleys to act as an aid to mariners and to let pirates know that a warship was nearby. The first lightship in Great Britain came about when Robert Hamblin obtained permission from King George II to fit out an appropriate vessel as an aid to navigation. The single-masted ship, christened the *Nore,* took station at the Nore Sandbank in the Thames River estuary in 1731. The *Nore* boasted two ship's lanterns hoisted up the mast high above the deck and spaced twelve feet apart from the cross arm. The lanterns burned oil with flat wicks. Although the log of the *Nore* contains entries detailing the "almost futile struggles to keep the lanterns lit" during storms, the lightship proved to be an immediate success and other lightship stations soon followed.[2]

Lightship development in the United States lagged behind that of England and Europe. The first recorded use of a lightship—then known as a "lightboat"—in the United States was at Willoughby Spit in Virginia in 1820. The seventy-ton lightship, however, could not withstand the exposed position, and the service moved it to Craney Island, near Norfolk, Virginia. In 1821, four more "inside" lightships—vessels that were located within the shores of the United States—were established in Chesapeake Bay. Sandy Hook, New Jersey, became the location for the first permanent "outside" lightship in 1823. The station's name changed to the Ambrose Channel Lightship on 1 December 1908.[3]

The use of lightships by the United States grew steadily. Although some of the vessels weighed more than one hundred tons, the term "lightboat" continued in use. When the craft finally earned the name lightship is unknown, but it was sometime after the U.S. Lighthouse Board took over the service from Stephen Pleasonton in 1852. In any case, the year 1837 saw twenty-six U.S. lightships in operation, and by 1852 the number had grown to forty-two. Interestingly, thirty-seven years later in 1889, the number had shrunk to twenty-four, and the ships were located mainly along the East Coast and the Great Lakes. Screwpile lighthouses replaced many inside lightships located in the Chesapeake Bay. The number, however, began to grow again as other outside lightship stations were added to the service's roles, including stations on the West Coast. By 1917, there were fifty-three lightship stations.[4]

While under the control of the fifth auditor, lightships suffered the same problems as did lighthouses. The main thrust of Pleasonton's policy for the lightships was economy, not purpose. The fixation on economy caused lightship development in this country to lag behind the progress made in Europe. A quick glance at what was wrong under the fifth auditor and the reason for the dismal showing of lightships in this country is necessary. Initially, little consideration went into suitable design and construction characteristics of the vessels. "Early light vessels were largely a product of opinion and arbitrary judgment on the part of builders who were often ignorant of the true purpose of the vessel or its harsh operating environment." The ships were wooden hulled and carried no spare equipment. The vessels were poor light platforms; the full body, shoal draft, and light displacement combined to cause undue rolling and quick motion. One lightship skipper complained that "her broad bluff bow is not at all calculated to resist the fury of the sea, which in some of the gales we experience in the winter season, break against us and over us with almost impending fury." Yet another captain grumbled that the lightship was "similar to a barrel," so that "she is constantly in motion, and when it is in any ways rough, she rolls and labors to such a degree as to heave the glass out of the lanterns, the beds out of the berths, tearing out the chain-plates, etc. and rendering her unsafe and uncomfortable."[5]

By 1842, the lightships in the United States tended to be from 40 to 230 tons burden and built entirely of wood, inadequately rigged for sail, and had no machinery-driven means of propulsion. If a storm drove a ship off station, the vessel had to sail to a port and then wait for the right weather to sail back to station. Worse, at this time there were no additional relief ships or tenders. If a lightship did go off station, it might be weeks or months before the ship returned, a dangerous situation for any navigator depending upon the light. The rebuttal of one local superintendent of lights to complaints about ships being off station was that when lightships were to be off station, a public notice was placed in the local newspaper—this did

This old print of artist Charles Nordhoff's etching "The South Shoal Lightship" shows the vessel in heavy weather.

little to help those at sea. Pleasonton had purchased a relief lightboat for Norfolk in 1837 that was still in operation during 1851. Clearly, one vessel was not adequate for the number of stations in operation.[6]

The lightships of the fifth auditor's period carried one or two lanterns, and the lanterns had either a compass or a common lamp with ten to twelve wicks protruding from it. The vessel at Hooper Straight, Maryland, for example, had a single lantern with eleven cylindrical wicks, while the lightship at Brandywine Shoal, Delaware, carried two oil lamps, each with twelve cylindrical wicks. Some early lightboats also carried fog signals of some type. The Diamond Shoals lightship mounted a six-pound cannon for fog.[7]

Another fault of the Pleasonton era was the practice of hiring farmers or landsmen as officers and crews on the ships. In turn, these men often either hired a stand-in or completely neglected their duties. For example, a fourteen-year-old African American youngster who had been left in charge of the lightship at Smith Point, Virginia, did not have the strength to hoist the lantern to the masthead.[8]

Under the new direction of the U.S. Lighthouse Board, lightships began to fare better. The board purchased spare equipment, including anchors and cables, while the lanterns sported new Argand lamps and parabolic reflectors. The board carried on the policy of trying to replace lightships with lighthouses. For example, in 1854 the first lightship completed under the U.S. Lighthouse Board's control took station at Minot's Ledge. It relieved an older lightship that had been on station since the first Minot's Ledge Light Station crashed into the sea in 1851. The lightship departed once the new light station went into operation in 1860.[9]

Lightships, like lighthouses, suffered during the Civil War. Confederates sank or appropriated the federal aids to navigation. The Smith Point vessel in Chesapeake Bay was sunk, as was the craft at Martin's Industry, the station for the Savannah River in Georgia. The 1862 report of the U.S. Lighthouse Board notes that all the lightships "from Cape Henry southwards, including the two in the Potomac River and those in Chesapeake Bay (except Hooper's Straits and Jane's Island), have been removed and sunk or destroyed by the insurgents."[10]

The U.S. Lighthouse Board moved quickly to replace the lightships. In 1862, the board purchased the *A. J. W. Applegarth*, a 203-ton brig, and rigged it for lightship duty at the Smith Point station. Other ships soon followed. The board requested army guards for the reestablished stations in case the rebels decided to capture the vessels.[11]

In the days when lightships were wooden and propelled by sail, it was often difficult for a mariner to recognize a light vessel from other sailing craft. Visual recognition of the lightships evolved through trial and error. The board experimented with various paint schemes and markings in the attempt to distinguish the ships from other merchant vessels and to provide the best contrast against the shoreline. Black was the color scheme of many of the early lightships. Combinations of red, black, gray, and various shades of yellow followed; a few even sported multicolored checks and stripes. By 1900, red and black became the predominant color, and the U.S. Coast Guard adopted in 1940 a standard color scheme for lightships: red hull with white lettering, white superstructure, and buff-colored stack, masts, lantern galleries, and ventilators. In January 1941, all U.S. lightships, except Ambrose Channel and Lake Huron, fell into line with the other ships. Ambrose converted from black to red in 1945, but Lake Huron remained black until discontinued in 1970.[12]

Until 1867, lightships had the station name painted on their sides, but no specifications existed for their marking or color scheme. In 1867, the U.S. Lighthouse Board began to number the lightships to keep better track of them no matter what

Charles Nordhoff captures the drama of serving on board a lightship in his etching titled "The Fog Bell."

station a vessel occupied. The numbers began in the north, with lightship number one being New South Shoal of Nantucket, Massachusetts. Eventually, numbering came to represent the chronological age of the lightship. From 1867 to 1913, lightships displayed their number on the stern and then on the bow and quarter. Often the station name was lengthy and, combined with the station number, required small lettering. To improve visibility of the lettering, lightships were directed to paint only one word for the station in large letters. For example, one station from 1894 to 1896 had a combination of numbers and lettering that spelled out "58 NANTUCKET NEW SOUTH SHOAL," which in 1913 was considerably shortened to "NANTUCKET."[13]

In 1880, the French began experiments on hulls that would remain stable in heavy seas. The British took these results and incorporated them into a ninety-foot iron ship with three watertight compartments below deck. Two years after the French began their experiments, the U.S. Lighthouse Board had the first U.S. iron light vessel, *No. 44*, built and sent to New Jersey. In 1891, the first three lightships with machinery propulsion came into service and were quickly ordered to the Great Lakes. A brief glimpse at these three lightships provides a good picture of how they performed under the U.S. Lighthouse Board, and the service the ships provided to the maritime community.[14]

No. 57, one of the first three lightships with machinery propulsion, sails from Gray's Reef Station in Lake Michigan in 1914. The lightship served for thirty-two years. No. 57 is displaying the triangular flag of the U.S. Lighthouse Service.

Lightships *No. 55, No. 56,* and *No. 57* began their duties on the Great Lakes on 22 October 1891, when they took up their respective stations on Simmons Reef, White Shoal, and Gray's Reef in northern Lake Michigan.[15] The ability of these unique steam-powered ships to move without sail caused the service to boast that they were "the only light-ships in any service to which this is possible."[16] The lightships mark the beginning of the continuous use of lightships on the Great Lakes. Earlier however, there had been a period from 1832 to 1851, when a lightship took station at the busy Straits of Mackinac, but the Waugoshance Lighthouse replaced the ship.[17]

The wooden steam screw lightships were built in Toledo, Ohio, by the Craig Ship Building Company for the total cost of $40,800, plus an additional $1,875 for installing steam windlasses and trysail masts. Their ribs were white oak and the fastenings were eight-inch, ⅝-inch square, iron spikes. Each ship was 102 feet, 8 inches in length, with a beam of 20 feet and a draft of 8 feet, 9 inches; the ships had a displacement of 130 tons. Each lightship had a high-pressure steam engine, with a fourteen-inch diameter cylinder, and a ten-foot-long marine fire box boiler five feet in diameter, with a working pressure of one hundred pounds of steam. The engines and boilers were all manufactured by the Eagle Iron Works of Detroit, Michigan.[18]

The illuminating apparatus on each ship consisted of two lanterns, each with three lamps burning mineral oil (kerosene); one lantern was on the foremast, the other on the mainmast. Lightship *No. 55*'s lanterns were a fixed red, *No. 56*'s a fixed white, and *No. 57*'s was white on the foremast and red on the mainmast. The three lamps of each lantern were suspended on a circular frame, which was hoisted up the mast. Each ship carried a bell weighing 33 ⅓ pounds and a 6-inch steam whistle for fog signals.[19] The normal complement of each of the three lightships was four officers and two crew members.

On 14 and 15 September 1891, the three lightships arrived at Detroit for examination by the service. Local U.S. inspectors examined the ships on 2 October, and, although many defects were uncovered, they passed this preliminary inspection. Next, sea trials took place on 5 and 6 October and the ships were found capable of making seven knots. The ships received ballast, underwent some modifications in the quarters of the officers and crew, then the lamps were fitted to the masts and the fog signals were adjusted.[20]

The tender *Dahlia* took the lightships in tow after they had reached Port Huron, Michigan, and before nightfall of 24 October, the three vessels were fast to their permanent moorings. Each mooring consisted of a "5-ton sinker with 15 fathoms of 2-inch chain on each." The ends of the chain were buoyed with spar buoys and were rigged to attach to the starboard bow chains of the light vessels. There was an adequate amount of chain to allow the lightships to ride out heavy weather. The permanent moorings had been placed in position by the tender *Warrington*.[21]

Orders to the Ninth Lighthouse District concerning the duties of the service were clear: "keep all lights burning and all light vessels and buoys in position, if possible, to close of navigation or as late as can be without injury or loss."[22] On 17 and 20 November, however, for some unknown reason, the three lightships departed their stations before the close of navigation and ran to Cheboygan, Michigan, their winter quarters. The Ninth Lighthouse District inspector promptly sent the ships back to their stations under the escort of the tender *Dahlia* on 23 November, and they remained on station until the close of navigation.[23]

The annual report of the service stated that the "officers and crew of these vessels, with one exception, were discharged for this dereliction of duty, and other men who have shown themselves more trustworthy, were put in their places."[24] The details of the case have been lost, but the dereliction of duty did help lead to reform. On 8 January 1892, the secretary of the treasury authorized the appointments of keepers and assistant keepers by the U.S. Lighthouse Board rather than by the collectors of customs.[25]

By the middle of the next navigation season, sailors plying the Great Lakes spoke "of these light-ships as being of the greatest assistance to them in navigating the Straits of Mackinac." The years passed with the lightships undergoing various modifications, but continuing to steam from winter quarters in the spring to their moorings in northern Lake Michigan and then returning to quarters at the close of navigation. In the end, primarily because of age, the ships became too costly to maintain.[26]

The first to go was Lightship *No. 55*. On 25 August 1920, the vessel was relieved of her duties on the Lansing Shoal Station and was sold on 15 February 1922 for $840. *No. 57* was the next to go in April 1924 and was "considered a barge for use in the Twelfth Lighthouse District as field equipment in connection with construction and repair." The remaining lightship of the original trio, *No. 56*, sold for $1,100 on 20 December 1928. Thus came the end of the first three machinery-powered lightships of the service. All in all, they had a long history of service to the maritime community of the Great Lakes. Lightship *No. 55* was thirty years old at the time of her retirement; *No. 57*, thirty-two years; and *No. 56*, thirty-seven years.[27]

It was not until the second half of the nineteenth century that marine architects began to look upon lightships as having a special duty and requiring a special design to carry out the mission. The bottom of the hull was flattened, and bilge keels installed to restrict the rolling about which the crewmen constantly complained. Naval architects also reduced the metacentric height so that the vessel would be steadier in the water.[28]

Wooden lightships in certain latitudes were subject to the marine worms that bored into the planks. In 1886, the U.S. Lighthouse Board recommended iron hulls for new lightships. There was, however, a strong debate about whether iron hulls could withstand the shocks that wooden ones could absorb. To gradually ease into all-metal hulls, in 1881 the board built composite hulls—part metal, part wooden skins. Lightship *No. 43*, for example, had a hull of metal plates sheathed with yellow pine planking. Another type of composite hull consisted of steel ribs and wooden skin and some metal plates. By the turn of the twentieth century, most recognized the superiority of metal hulls, although the board constructed *No. 74* entirely of wood in 1902. Despite the presence of the borers, many lightships led long lives. Light vessel *No. 23,* for example, lasted sixty-eight years, while *No. 29* remained in service for fifty-two years.[29]

Along with changes in the design and propulsion of lightships came better living conditions aboard the vessels. Lightships were originally single decked, with the crew living below decks. The first changes for the crew involved building up the forecastle and decking it over so the crew could reside forward. Eventually, the crew berthed above the water line with enough room that officers could have a room to themselves, while the crew had two-man rooms.

Ground tackle (anchors and chains) improved. Lightship crews were expected to keep station in all weather. In the days of the fifth auditor, however, many ships had only light twelve-hundred-pound anchors and no independent means of propulsion; it is little wonder that many ships were blown off station. By the years of the U.S. Lighthouse Board, the idea of keeping station was becoming more realistic. As mentioned, lightships *Nos., 55, 56,* and *57* in the Great Lakes had steam power and moored to five-ton sinkers. By 1917, ground tackle weighing fourteen tons held the Diamond Shoal lightship on station. The heavier hooks and machinery propulsion made station keeping more feasible, although not completely infallible. A brief glance at the station at Cape Hatteras gives a good overview of the changes in lightship station keeping.[30]

As early as 1806, Congress authorized a survey of the coast of North Carolina to mark the shoals at Cape Hatteras, Cape Lookout, and Frying Pan Shoals. Congress noted "it is supposed there is no part of the American coast where vessels are more exposed to shipwreck, than they are in passing along the shores of North Carolina, in the neighborhood of these shoals. . . ." The survey was also to determine if it was feasible to build a lighthouse on the tip of Cape Hatteras Shoals. The following January saw the completion of the survey party's work. The report stated it would not be advisable to build a lighthouse on the shoals because the bottom consisted "of loose and shifting sand" and no lighthouse could withstand the constant blast of the ocean at the shoals. Thirteen years later, in 1819, the U.S. Senate requested yet another survey. President James Monroe instructed the navy to undertake the project. The survey recommended that lightships be placed off Cape Hatteras, Cape Lookout, and Cape Fear.[31]

Congress appropriated money for two lightships, one of which was to be stationed at Cape Hatteras. Pleasonton had specifications for the ships drawn up and, in May 1823, Henry Ekford of New York received the contract for the two vessels. Construction went well, and in April 1824, Pleasonton notified Capt. Christian Erickson of Philadelphia that he was now master of the Cape Hatteras lightship, with a salary of $800 a year. Erickson was to report to the ship on 10 April 1824. The vessel displaced more than three hundred tons and had two lights, one sixty feet high and the other forty-five feet. Erickson's crew was to consist of a mate and four or five seamen.[32]

In June, the Cape Hatteras lightship was ready for sea. Supplied with oil, wood, and water, the ship departed around the middle of June and arrived on her station before the end of the month. Accompanying the lightship was Capt. Jesse Elliott, who was instrumental in getting Congress to pass the law for the establishment of the lightship at Cape Hatteras. Captain Elliott had impressed Pleasonton with his help during the construction of the ship and his offer to accompany the ship on the way to the station. Captain Elliott, however, disapproved of Erickson's "intemperate conduct." As with much of the history of the service, the details of the "intemperate conduct" have been lost. In any case, Elliott wrote of his feelings to Pleasonton. The fifth auditor, again showing his willingness to listen to someone who appeared to be knowledgeable in nautical affairs, recommended the removal of Erickson and the appointment of Capt. Lief Holden. On 31 July 1824, the president followed up, appointing Holden master of the Cape Hatteras lightship.[33]

The light for the vessel was designed by Commo. James Barron. Another of the many controversial and colorful characters that continually seem to inhabit the history of the service, Barron had killed the naval hero Stephen Decatur in a duel, and he earlier had surrendered the frigate *Chesapeake* to the British *Leopard*. As a result of his duel, Barron never received another naval assignment and turned to devising a lightship lighting apparatus. The acceptance of Barron's device may have come about because Captain Elliott had served under him on the *Chesapeake,* and Elliott had the ear of Pleasonton. The features of Barron's design is not known, but in any case, Pleasonton bought lamps at $550 for three lightships.[34]

The lightship was positioned thirteen miles east-southeast of the Cape Hatteras lighthouse. A severe storm in February 1825 parted the vessel's moorings, driving it many miles out to sea. On 12 February, the ship limped into Norfolk, Virginia, for repairs. The repair work creaked along, and Pleasonton felt the ship not ready for station until 15 December. A six-pound cannon used as a fog signaling device became a part of the lightship's inventory during this long inport period.

In May 1826, the lightship again parted her mooring and once again ran to port. Pleasonton's reaction to the loss of the ship's anchor is one of the best examples of the fifth auditor's excessive focus on money, while forgetting the need to provide dependable aids to navigation. When Pleasonton learned of the loss of the anchor, he ordered ships to search for the hook and offered a five hundred-dollar reward to anyone locating it. At the same time, he ordered the ship to remain inport while the search ensued. Toward the end of June, almost a month after the ship came into harbor, Pleasonton finally decided to let out a contract for a new anchor. After manufacturing and testing, the ship was off station about five months. As F. Ross Holland has aptly put it, "One wonders how many ships went aground looking for the lightship that wasn't there," all because Pleasonton could only think of money.[35]

In August 1827, the lightship again broke her moorings and slammed ashore six miles south of Ocracoke Inlet. The vessel was not salvageable and the collector of customs at Ocracoke received orders to break up the ship.

In 1889, Congress approved an appropriation of $200,000 for the construction of a lighthouse on Outer Diamond Shoal, off Cape Hatteras. The efforts to build the light were unsuccessful, and the remaining funds were diverted to construct lightship *No. 69* and reestablish a lightship at the station now called Diamond Shoal. After more than a seventy-year break, a lightship would now be back near the old Cape Hatteras location. By this time, of course, the U.S. Lighthouse Board was in control, and lightship design had improved considerably. The new lightship, built by Bath Iron Works of Bath, Maine, cost $200,000. The vessel had a composite hull—a steel frame with wood bottom and steel-plated topside—and two masts with lantern galleries on each. She was steam powered, with a single four-bladed propeller, eight feet in diameter. She could make 8 ½ knots under steam power but still had sails. Lightship *No. 69* was 122 feet, 10 inches in length, with a beam of 29 feet, 6 inches and a draft of 13 feet, 6 inches and displaced 590 tons. Her lighting apparatus consisted of a cluster of three one-hundred-candle-power electric lens lanterns permanently mounted in a gallery at each masthead. She also carried a twelve-inch steam-chime whistle and a hand-operated one thousand-pound bell as fog signaling devices.

Lightships *No. 69* and *No. 71* alternately relieved each other on the Diamond Shoal station until 1901, when another vessel was assigned. A lightship remained on station, except during World War II, until disestablished in 1966. Beginning in 1824, the vessels assigned to this dangerous station were blown adrift or dragged off station twenty-two times in severe weather, with one ship going ashore and on two other occasions sustaining severe damage. On 6 August 1918, the vessel assigned to Diamond Shoal, *No. 71,* had the dubious distinction of being sunk by surface gunfire from the German submarine, *U-104.* The submarine's captain allowed the crew to take to the lifeboats before opening fire and the doomed lightship's crew made it safely ashore.

Most of the terrifying moments on the station, however, were the results of the infamous weather of the area. The crew of *No. 105,* for example, during 15 and 16 September 1933, battled a hurricane. Even with the engine at full speed, she was dragged five miles into the breakers on the southwest portion of Diamond Shoal. The boats, engine room ventilators, and radio antennas were carried away. In the fire room, water rose above the floor plates. Then the winds shifted and *No. 105* drifted from the breakers to sixty miles east-northeast of Cape Hatteras. Despite all the pounding, the vessel steamed into Portsmouth, Virginia, on 18 September. President Franklin D. Roosevelt commended the officers and crew on their seamanship. Three years later, on 17 and 18 September 1936, another hurricane, packing winds of one hundred miles per hour, dragged the lightship one and a half miles off station.[36]

The West Coast of the United States had fewer lightships than the East Coast, but some locations were just as difficult. The first lightship station along the U.S. Pacific shoreline—off Oregon's coast—marked the approach to the entrance of the important Columbia River and is a good example of the challenges of the West Coast.

The first U.S. lightship on the West Coast, *No. 50,* arrived at its station, 4.4 miles southwest of Cape Disappointment, Washington, in 1892. Built by the Union Iron Works, San Francisco, for $61,150, it slipped down the ways in 1892. Light vessel *No. 50* was 120 feet, 10 inches in length, with a beam of 26 feet, 9 inches and a draft of 11 feet, 3 inches. The vessel displaced 470 tons. She had a composite hull with steel ribs, was planked with Puget Sound pine, and was sheathed with white oak and fastened with galvanized iron bolts. She was sail powered and schooner rigged. The vessel carried a 5,000-pound mushroom anchor.[37]

The lightship carried out her duties routinely until a severe gale whipped the Oregon coast on 28 November 1899. The *No. 50* fought at the end of her anchor chain, but the seventy-four mile an hour winds and seas proved too much and that night she broke free of her tether. The skipper managed to work *No. 50* some twenty-five miles offshore and waited out the night before attempting any passage to safety.

The lighthouse inspector dispatched two tugs and the district tender to assist *No. 50.* Although a tug managed to get a line on her, the towing hawser snapped. The tender *Manzanita* worked a line to the lightship, but the line parted and fouled the tender's screw. Now two ships were in peril. The second tug managed to make fast to *No. 50* and began the always tricky and dangerous job of bringing a vessel over a turbulent bar. When the hawser parted again, the master, Joseph H. Harriman, made a quick and difficult decision: he would try to beach the vessel rather than see her pounded to pieces on the rocks. Harriman pointed the helpless lightship toward a sandy shore and skillfully maneuvered *No. 50* onto the beach. He beached her stern first, thus helping prevent the vessel from rolling. The crew then safely abandoned ship.

Efforts to refloat *No. 50* proved unsuccessful. In July 1900, the U.S. Lighthouse Board took the highly unusual action of moving the ship *overland* and then relaunching her in the shelter of Baker's Bay. On 14 February 1901, the strange task of bringing a light vessel over seven hundred yards of land began. The first order of business consisted of removing the sand from the stranded lightship. Then, rails were laid from the ship over the seven hundred yards to Baker's Bay. Next, horses and winches slowly tugged and pulled the ship over land to the bay.[38]

The next light vessel to occupy the Columbia River station for a long period, *No. 88,* demonstrates the difference between wood-and-sail craft and those of steam and steel. Lightship *No. 88* was built in 1907 by the New York Shipbuilding Company for $90,000 at Camden, New Jersey. She was steel hulled, with a wooden pilothouse and deckhouses. She had two steel masts, a stack amidships, and a steam-compound reciprocating engine. Her single screw was seven feet, nine inches in diameter and the vessel had a designed maximum speed of ten knots. *No. 88*'s original design called for her to be rigged for sail.

The illumination for *No. 88* was a cluster of three oil lanterns raised to each masthead. In 1927, the vessel had 375-mm electric lens lanterns for each masthead. The lightship's original fog signals were a twelve-inch steam-chime whistle and a hand-operated bell.[39]

Lightship *No. 88* began her thirty-year assignment on the Columbia River station in 1909 after a long journey from New York in a convoy with two other lightships and three tenders. Five years after first taking station, a dramatic incident proved the worth of a steam-powered, iron-hulled vessel. On 2 January 1914, a vicious storm lashed Oregon's coast. Into the teeth of the storm, *No. 88* steamed ahead for eleven hours. A boarding sea slammed across the lightship. In the pilot house, the force of the sea sheared off the steering compass, bent the steering wheel shaft and carried away the bridge binnacle. The waves also swept away the ventila-

The first lightship on the West Coast, No. 50 *was blown ashore near the Columbia River in 1899. A year later it was transported seven hundred yards by rail in order to be refloated in a sheltered bay. The lightship served from 1892 to 1909 and then was laid up until 1915, when it was condemned.*

tor and galley stack and damaged the engine room skylight, flooding the room. The sea was so strong that it smashed portholes in the after house and carried away deck stanchions. Two crewmen received injuries from flying glass. The skipper matter-of-factly noted: "The damage is not such as we cannot wait until the vessel comes in for her annual repairs." Such a report would probably not have been penned if the ship had been a wood-and-sail light vessel.[40]

Lightship *No. 88* went on to continue her duties on the Columbia River station until 1939 and then shifted to other stations, mostly along the Washington coast. At the age of 53, the lightship retired from duty in 1960. Old *No. 88* became a floating exhibit at the Columbia River Museum from 1963 to 1979 and then was sold. In 1984, the former light vessel was converted to a half brig and renamed the *Belle Blonde;* she sailed to Quebec in 1984 for charter use.[41]

DUTY ABOARD LIGHTSHIPS consisted of a strange mixture of danger, uncomfortable conditions, and boredom. One writer said that an old salt once remarked the "lonliest object" he had ever seen was the South Shoal lightship.[42] Lightship duty was arguably the most dangerous in the Lighthouse Service. Real danger came generally from two sources. The first was the weather. In November 1913, for example, Lightship *No. 82* on station some thirteen miles southwest of Buffalo, New York, disappeared in a storm, with a crew of six. Six months later, divers located the submerged wreck two miles from her station. A diver reached the wreck sixty-three feet below the surface of Lake Erie's waters and reported that the storm had snapped her cables and battered her superstructure. Only one body was found a year later some thirteen miles from the wreck site.[43]

Lightship *No. 37* foundered at her moorings at Five Fathom Bank, off Delaware Bay, on 23 and 24 August 1893. The crew began to notice that the ship, battered by wave after huge wave, developed a port list, which became progressively worse after each wave. The seas had already carried away the vessel's lifeboats. At 0147 hours, 24 August, four successive seas struck *No. 37* broadside, sinking her immediately. The assistant engineer, trapped below, somehow managed to work his way free and fought his way to the surface, where he grabbed some debris from the lightship. Only he and one other crewman survived out of the six man crew.[44]

When the gale hit *No. 88* at the Columbia River on 2 January 1914 and swept the decks and pilot house, Seaman H. K. Hansen on duty in the wheelhouse was swept off his feet by the boarding sea and carried along with the wreckage toward the side. Just before being swept overboard, Hansen managed to reach out and grab the rail.[45] Another West Coast station, Swiftsure Bank, at the mouth of the Strait of Juan de Fuca, also suffered through a severe storm. In December 1932, the skipper of the lightship recorded that, during a storm, the "water rushed into the pilot house with terrific impact hitting seaman Harris who was then on watch. Threw him against the steering wheel with full force . . . Seaman Harris was knocked unconscious and suffered a broken arm and had many cuts on top of his head, forehead

A lightship's mate checks the lanterns some forty feet above the deck.

and arms. He also suffered a badly wrenched back. Water had been breathed into his lungs from lying in about a foot of water. . . ." Harris was found by his shipmates and was revived and bandaged.[46]

The second greatest danger aboard a lightship was from collision. Many collisions were minor mishaps with barges and fishing vessels, but others were serious. The lightships on the Scotland, New Jersey, station were struck six times from 1892 to 1905, with four of the collisions coming in the period from 1903 to 1905. On 24 April 1919, a Standard Oil Company barge struck *No. 51* while on the Cornfield Point station off Connecticut. The ship sank within eight minutes, but no lives were lost. At least 150 collisions involving lightships are in the records of the service. This does not count the bumps, sideswipes, and near misses.[47]

On 6 January 1934, the lookout of *No. 34*, occupying the Nantucket Shoals, Massachusetts, station, saw the SS *Washington* looming out of the heavy fog. The ship sideswiped the lightship and caused the radio antenna on the yard to be carried away and some hull plates to buckle. A little more than four months later, on 15 May, the large passenger liner SS *Olympic*, sister ship of the ill-fated *Titanic*, loomed out of another thick fog and struck *No. 34*. She sank beneath the waves, carrying four of the eleven-man crew down with her. The *Olympic* picked up seven crewmen, but three of the survivors later died of injuries and exposure. The British government paid for building the replacement *No. 112* as part of the reparation for the incident.[48]

Lightship, *No. 6,* occupying another Massachusetts station, Cross Rip, on 4 February 1918 drifted slowly eastward in heavy ice. A keeper of the Great Point light reported the lightship still drifting in the ice the next day. Lighthouse Service and U.S. Navy ships began a search, but no trace of the vessel or the six men was found. Over the years, various artifacts brought in from the sea are thought to be parts of the *No. 6.* The latest, a lighthouse bell, was recovered in 1987.[49]

Most incidents were not as destructive as with *No. 34* or *No. 6.* Often, the worst was a sideswipe, being blown off station, or the uncomfortable times of riding out a severe gale. But the potential danger was there in the minds of the crew. The author can remember shipmates who had served on lightships saying a number of crewmen were convinced that some ships used the lightship's radio beacons to home in on during fog and worried that if someone was asleep on the bridge of the inbound ship, the lightship stood a chance of being cut in half.

The daily life of early lightship sailors seesawed between boredom and discomfort. One crewman remarked, "If it were not for the disgrace it would bring my family, I'd rather go to State's prison."[50] A New England sailor recalled, "I had an idea it must be terrible lonesome aboard one of them critters, but I had to go and find out and I did. . . . I have seen boys or young men come on there and after a few days when the newness had worn off they would be thinking of home and mother. I would find them pacing the deck and looking inshore. . . . When the tenders would come out to water us up or to coal the ship, they would bring our mail, and then those boys—Barnum never put on a better show."[51] The early lightships, as has been pointed out, were difficult to ride in bad weather. Even old salts could become violently seasick on lightships. On calm days came the tedium. Taking care of the light and keeping the ship squared away took only a few hours of the day. Very few of the lightshipmen were readers and some passed away the long hours making rattan baskets to sell ashore. The food was good but changed little. The favorite dish was scouse, a "wonderful commingling of salt beef, potatoes, and onions." Two watches were set—the captain's and the mate's. Duty stretches were long, with most crewmen spending at least eight months on the ship.[52]

Long stretches of isolated, lonely duty can cause strange behavior in people. On one vessel, the captain and mate got on each other's nerves so badly that they could not eat their meals unless a curtain divided them. Even in modern times, with the length of duty aboard a vessel reduced, situations could be difficult. A quiet crewman on the Umatilla station, off Washington State's coast, became obsessed with reading horror and murder mysteries, but only when fog settled in around the lightship. Not only would he read the frightening tales, he would discuss them at great length with his shipmates. He quickly earned the sobriquet of "Mr. Murder." Once the fog would lift, he would resume his normal, quiet persona. During one long foggy stretch, the crew decided it had had enough. The sailors rigged a ghost made of the proverbial sheet and attached it to a line running from the foremast to the bridge area; they waited for the required "dark and foggy night." Mr. Murder came on watch by himself at midnight on such a night. He took up his station on the wing of the bridge. Suddenly a wraith appeared before him. He shook it off, but the apparition again floated toward him. He screamed and fled from the bridge to tell his shipmates what happened. Supposedly, the man never read another horror story aboard the lightship.[53]

Although the primary duty of the crewmen was to keep the lightship on station and maintain its light, the tradition of the sea in helping others in distress was carried out by the sailors about the light vessels. Two unusual occurrences happened in the Pacific.

The lookout on *No. 83*, occupying the Blunt's Reef, California, station was startled at 0130 hours on 15 June 1916 to hear the sound of human voices drifting to him through the thick fog surrounding the lightship. Dimly through the mist, the now wide-awake sailor made out a lifeboat cutting through the fog. He rang the emergency bell and soon the lifeboat, filled with women and children, came alongside. This was the first of many boatloads from the steamer *Bear*, which had stranded nearby. Eventually, 155 men, women, and children were crowded aboard *No. 83*. The survivors were later transferred to other nearby vessels.

Lightship *No. 70*, occupying the San Francisco station, also proved a haven that may stretch the limits of what is considered rescue. On 2 May 1901, the crew found that their ship was a home to many land birds, such as owls, humming birds, and cranes. Apparently they had become lost and confused during a large forest fire not far from the ship's location. Once the smoke cleared, the birds departed.[54]

A series of incidents not stretching the limits of the word rescue happened in 1916. Prior to the entry of the United States in World War I, a German submarine raider visited Newport, Rhode Island. The submarine got under way and eventually took up station near the location of the Nantucket lightship and waited for its prey. Before the alarm spread, the sub sank a number of merchantmen and brought the survivors to the lightship. According to Commissioner of Lighthouses George Putnam, at "one time there were 115 shipwrecked men aboard the lightship, and 19 ships' boats were trailing on a line astern. . . . It is certain that many of these seamen would have lost their lives had it not been for the haven provided by the lightship."[55]

When technology perfected a device to help the mariner, that device soon came aboard the lightship. In 1906, for example, the submarine signal was introduced. This device hung over the side of a light vessel and produced an underwater sound by a bell-shaped casting with an internal striker operated either mechanically or by compressed gas. Each lightship transmitted a distinctive code and could be detected by another vessel with sensors forward on each side of its hull and through headphones. The device had a ten-mile range, but some reached twenty-two miles. A submarine oscillator came into service around 1923 and signals reached ranges of up to forty miles. The radio beacon replaced the submarine signal; this navigational aid was first introduced to U.S. lightships on the Ambrose station in 1921.[56]

A little more than two years after the U.S. Coast Guard took over the U.S. Lighthouse Service, the United States entered World War II. Many, but not all, lightships were withdrawn from their stations and replaced by buoys. The remaining lightships were armed; many were designated submarine net tenders and performed ship recognition duties near major harbors. After the end of the war, the lightships shed their guns and again resumed their duties.

The U.S. Coast Guard continued to follow the Lighthouse Service policy of replacing lightships. Because of new technology, the lightships began to disappear. Interestingly, this happened just as conditions improved greatly aboard lightships. The hull design, for example, helped to improve stability. Radio and television, plus duty stretches usually of not more than thirty days, helped to fight the feeling of complete isolation. The blast of the fog horn and the smell of diesel fuel, however, could still make some crewmen uncomfortable.[57]

In 1957, the means of finally replacing costly lightships became available. The U.S. Coast Guard began feasibility studies on using offshore light structures to replace some lightships. The service noted that a lightship, plus a relief vessel, required 1.32 vessels, which translated into a cost of $1,320,000 per station. A template of the type

Large Navigational Buoys (LNB) helped spell the end of lightships.

of structure was based on the offshore oil drilling platforms in the Gulf of Mexico. In 1961, a contract was awarded for the first two structures to replace the Buzzard's Bay, Massachusetts, and the Brenton Reef, Rhode Island, lightships.[58]

Commissioned on 1 November 1961, Buzzard's Bay received the first structure. The primary contractor was the Perini Corporation of Boston. The structure stands in sixty-one feet of water. Four legs support the structure, and it is eighty-two and one-half feet in overall height. Inside the main legs, 30-inch steel pipe piles anchor in bedrock at approximately 268 feet below mean low water. The legs support a seventy-by-seventy-foot superstructure, which provides living spaces for a crew of five, machinery areas, and fuel storage. A helicopter landing pad is on the top deck, and a fog horn is on a corner below the landing pad. The light tower at one end of the main deck is a six-foot diameter steel column with an internal spiral stairway leading to the lantern room. Mounted on the roof above the lantern is a radio beacon antenna. The superstructure is red. The blast of the fog signal carries for five miles, the light can be seen for sixteen miles, and the radio beacon's range is seventy-five miles. By 1995, all the offshore structures were automated.[59]

In depths of one hundred feet or more, or in areas where the bottom is not suitable for the fixed offshore structures, the U.S. Coast Guard developed the large navigational buoy, or LNB. In June 1966, the U.S. Coast Guard awarded the Convair Division of General Dynamics the contract to construct the prototype buoy. The buoy has a discus-shaped hull, forty feet in diameter, by seven and one-half feet deep, and is built of steel with all-welded construction. The main body of the buoy has many watertight compartments made up of a series of bulkheads. In the center of the buoy is a thirty-five-foot mast-tower, with a thirty-five-foot radio beacon whip antenna mounted on top. The buoy displaces 60 tons unballasted and 104 tons ballasted. Draft of the buoy at full ballast is three and three-quarters feet.

The LNB was first delivered to the U.S. Coast Guard on 26 May 1967 and shortly after that moored in fifty feet of water on the site of the former Scotland Lightship at the southern entrance to New York harbor. Navigational aids on the buoy consist of lights, a sound signal, and a radio beacon. They are controlled either by automatic equipment within the buoy or remotely from a U.S. Coast Guard station via a UHF-FM radio circuit. The shore station equipment monitors the buoy's operations. In January 1968, equipment to measure some oceanographic and meteorological parameters were added to the buoy.[60] By 1995, electronics changed navigation so that only one LNB remained on station at Nantucket Shoals, and it "will probably be removed and replaced by a smaller buoy late 1996 or early 1997" when the global positioning system (GPS) is accepted as a worldwide navigation system.[61]

THE END OF LIGHTSHIPS came on 29 March 1985, as it was destined, when *Nantucket I* was decommissioned. Thus ended an era of maritime history. At 144 years of age, the Ambrose Channel station was the longest-maintained U.S. light station. While lightship duty was dangerous, monotonous, and lonely, many sailors

On 1 November 1961 the Texas Tower–type light station first began to replace lightships at selected locations.

would have no other. When the *No. 67* retired from lightship duty in 1930, Eric H. Lindman, who had served aboard the *No. 67* as both mate and master since 1915, "had tears in his eyes." Lindman "had always spoken of his command as if it were the greatest little ship in the world." Perhaps a fitting epitaph for these small vessels that did so much to help other vessels safely make port is in the final message of *Nantucket I:* "An important part of Coast Guard history ended today. We must now look somewhere else to find the stuff that sea stories are made of."[62]

The Black Fleet

PART OF THE U.S. LIGHTHOUSE SERVICE'S FLEET consisted of ships that are often overlooked by those who prepare coffee table books on vessels of all types: the unglamorous tenders. Compared to sleek destroyers and cruisers of the navy or graceful passenger liners, tenders are like tug boats among racing yachts. Tenders are the ships that do the servicing and repairing of aids to navigation and other hard maritime work. A secretary of commerce once wrote that tenders are "vessels whose duty it is to go where no other vessels are allowed to go, and who, through storm, darkness and sunshine, do their work. . . ."[1] If it were not for the hard and dirty work of the tenders, many more glamorous ships would have difficulty in reaching a safe berth.

Under fifth auditor Stephen Pleasonton, private contractors did the necessary maintenance on lighthouses and buoys. In 1840, the U.S. Lighthouse Service acquired its first lighthouse tender by purchasing the ten-year-old U.S. Revenue Marine cutter *Rush*, which was beset by ice in New York harbor and then abandoned. For the next twenty years the *Rush* tended buoys in New York harbor. The lighthouse service, however, continued to contract for needed buoy work in other locations.[2]

An overview of the lighthouse tenders on the Great Lakes in the nineteenth century is illustrative of the history of these ships in one specific region. In May 1856, the U.S. Lighthouse Board purchased the 120-ton schooner *Challenge* for $6,250 to use as a supply vessel and a lighthouse tender for the Tenth Lighthouse District. On 29 April 1857, her name was changed to the more appropriate *Lamplighter*.[3] The *Lamplighter* was the first federally owned lighthouse tender on the Great Lakes.[4]

Less than five months later, on 11 September 1857, the *Lamplighter* wrecked off Lake Superior's Isle Royale. An investigation by the Eleventh Lighthouse District inspector reported the wreck was caused by the "neglect of duty of Capt. [J. W.]

Taylor, Master of Supply Vessel." The tug *Oswego,* dispatched from Detroit, started the tow of the *Lamplighter* to Detroit on 13 November.[5]

After repairs totaling $1,000 the *Lamplighter* again set out on her rounds. Illustrating the hazards of tenders, the *Lamplighter* once again ran aground, this time on a reef at Mackinaw, Michigan, in September 1859. The tender's repairs in April 1861 came to $150, plus a $125 fee for towing the *Lamplighter* off the reef.[6]

A year later, on 18 April 1862, Col. J. D. Graham, engineer for both the Tenth and Eleventh Lighthouse Districts, estimated an outlay of $4,000 for rebuilding the *Lamplighter,* but he recommended the U.S. Lighthouse Board sell the tender and purchase or charter another ship. On 6 May 1862, Colonel Graham followed up his recommendation with another letter "reporting [the] result of [his] efforts to procure a vessel for use as Tender for 11th District and reporting certain propositions to charter."[7] The same day, F. W. Myers of Buffalo, New York, in a letter to the U.S. Lighthouse Board, offered "to contract for delivering supplies & tending buoys in the 10th & 11th Dists. for $6,500 per annum."[8] The board decided to await "the promised further reply of Col. Graham on this subject," before making their decision.[9] Colonel Graham, however, restated his recommendation to sell the *Lamplighter* and suggested the chartering of the schooner *Dream.*[10]

The U.S. Lighthouse Board followed Colonel Graham's recommendation and chartered the topsail schooner *Dream* in May 1862. Built in 1859 and made of oak, the schooner was ninety-two feet in length, had an eighteen-and-a-half-foot beam, displaced eighty tons and, when loaded, drew only six feet, two inches. The *Dream* worked buoys for about six months, and her owners received $300 per month, "exclusive of the pay and subsistence of the Officers and crew."[11]

The same year in which the U.S. Lighthouse Board acquired the *Lamplighter,* it also obtained another schooner. Purchased for $6,000 in December 1856, the *Skylark* came down the ways in 1854, displacing 146 tons. She received the new name, *Watchful,* on 29 April 1857. The tender served in the Tenth Lighthouse District for ten years and later was disposed of at public auction for $3,050. Hastening the end of her days was the appearance of the first steam lighthouse tender on the Great Lakes on 16 October 1867.[12]

The eighty-foot *Belle* was the last sailing vessel operated by the U.S. Lighthouse Board as a tender on the Great Lakes. Built in 1861 at St. Joseph, Michigan (originally as the *Belle Stevens*), the schooner measured eighty-eight feet in length, twenty-one feet, three inches in the beam, and drew a shallow six feet when empty. Her hull was of oak and was iron fastened.[13]

The U.S. Lighthouse Board purchased the ship in 1863, renamed her *Belle,* and assigned her to the Eleventh District as a tender. At the time, the *Belle* was responsible for all aids to navigation above Detroit, which included those at Lakes St. Clair, Huron, Michigan, and Superior, plus the straits connecting them. In 1864, the district had eighty-five major aids to navigation consisting of forty-eight lights and thirty-seven buoys. The *Belle* was kept busy for "many of the buoys . . . [were] dragged from their stations by passing tow-boats."

Over the next three years, the *Belle* transported men and equipment to work on lighthouses, while continuing to work buoys. One report described the district buoyage as having been "well cared for, and few, if any complaints have been made in this respect." By 1867, the number of aids to navigation in the district had grown to sixty-three lighthouses and lighted beacons and eighty buoys. The U.S. Lighthouse Board then reversed themselves, reporting that the *Belle* had only "indifferently" performed her duties, since the sailing tender "could not possibly make more than one tour during the same season, and could spare but little time to devote to the buoyage of the district." One historian notes this sudden reversal is probably attributed to the "appearance of the first steam-powered lighthouse tender, the *Haze,* on the Great Lakes. . . ."

During the massive engineering project that went into building the Spectacle Reef lighthouse, the *Belle* was engaged in transporting material to the construction site. From 1869 to at least 1872, the *Belle* continued to work at Spectacle Reef, in addition to carrying out other duties. During the 1871 season, the tender served as the quarters for the work crew of the enormous construction effort.

On 10 November 1873, the district engineer of the Eleventh Lighthouse District notified the U.S. Lighthouse Board that the *Belle* had run aground and was damaged badly. The U.S. Lighthouse Board ordered her, on 10 July 1874, to be "sold at auction." The board received $470, and in May 1875 the last sailing lighthouse tender on the Great Lakes passed into history.

The first steam-powered tender on the Great Lakes, the *Haze,* "relieved two sailing vessels, performing all their duties and having enough time in excess to enable her to devote a considerable portion to duties not heretofore performed by sailing vessels."[14] The U.S. Lighthouse Board purchased the propeller *Haze* on 6 June 1867, from T. S. Winslow of New York City for $27,000. The steamer, built in 1867 by Henry Mallory of Mystic, Connecticut, measured 124 feet overall, 23 feet in beam, drew 8 feet, and displaced 274 tons. Her single expansion steam engine rated 292 horsepower.[15]

The steam-powered tender saw service in both the Tenth and Eleventh Lighthouse Districts, and her duties took her to nearly all corners of the Great Lakes. The annual report of the U.S. Lighthouse Board notes one of the reasons for such an underway record: "This vessel also carries such freight and parties for repairing, rebuilding, etc., of light-houses, before and after delivering supplies, as time and other duties will allow."[16] Within just three years, however, the U.S. Lighthouse Board stated that the *Haze* had "deteriorated to such an extent" that it recommended "she be rebuilt." Furthermore, the major reconstruction was needed because the *Haze* had not been "built for the lighthouse service, but was purchased in an emergency."[17]

Congress appropriated $30,000 on 3 March 1875 resulting in the *Haze* being "so enlarged and rebuilt at Buffalo, N.Y., in 1876, as to be practically a new vessel."[18] The rebuilt ship continued on into the twentieth century, still tending lighthouses and buoys. As late as 1894, the *Haze* continued to steam with only minor repairs needed and covered "some 8,860 miles."[19] On 15 March 1905, however, the tender was laid up at the service's depot at Tompkinsville, New York, and her officers and crew transferred to the *Crocus*. The next year, the *Haze* sold for $800.[20]

The single-screw steamer *Warrington*, purchased in 1870, succeeded the *Haze*. The *Warrington* continued in service until 1911. The *Dahlia* (1874), in turn, followed the *Warrington*. The *Dahlia* was the first iron hulled tender built from the hull up for use upon the Lakes. From 1865 onward, all tenders took their names from trees, shrubs, or flowers. The *Dahlia* was the first tender on the inland seas to have a botanical name. *Dahlia*'s career stretched over a thirty-five year period until 17 February 1909. The tender *Marigold* (1890) lasted well into the U.S. Coast Guard–era, ending her service in 1946. The last Great Lakes tender constructed during the nineteenth century was the *Amaranth*, launched on 18 December 1891 at Cleveland, Ohio. The smallest, at thirty feet in length—with a seven-foot beam, a draft of only three feet, and powered by a twenty horsepower engine—the *Lotus* served from 1880 until 1901.[21]

In general, the U.S. Lighthouse Service designed one-of-a-kind tenders for use in a particular district. The history of the *Shubrick*, the first steam lighthouse tender on the West Coast and the first designed by the U.S. Lighthouse Board, is an excellent illustration of the early steam ships used by the U.S. Lighthouse Service and some of the ships' missions.

Until 1852, the entire stretch of the U.S. Pacific coastline had no lighthouses or buoys to help mariners. The lack of aids to navigation was not a strong issue until after the settlement of the Oregon question in 1846 and after the gold discovery of 1849 in California. These two events caused a flood of emigrants, and thousands of people streamed to the West Coast. The large influx of emigrants also caused maritime commerce to increase. On 31 August 1852, Congress authorized $1,000 "for large buoys to be placed on sunken rocks in the Bay of San Francisco, under the direction of the Superintendent of Coast Survey."[22]

The buoys were nothing more than temporary wooden spars, and their upkeep was foisted off onto an already over-worked Coast Survey. Congress reluctantly authorized $60,000 on 18 August 1856 to build a lighthouse tender for the West Coast and to establish the first permanent buoys in those waters.[23]

The Philadelphia Navy Shipyard had just completed the construction of the frigate USS *Wabash*, and a great deal of material was left over. The U.S. Lighthouse Board, never slow in taking advantage of what they perceived as a bargain, used the remnants of the navy's ship to build the lighthouse tender *Shubrick*, named appropriately enough after the first head of the U.S. Lighthouse Board, Commo. W. B. Shubrick, U.S. Navy. Construction began on 25 November 1857.[24]

On 27 May 1858 the Shubrick became the first steam tender on the West Coast, and the first built from the keel up by the U.S. Lighthouse Board. The tender was sold in 1886.

The *Shubrick* was the first steam tender of the service and the first built from the keel up under the direction of the U.S. Lighthouse Board. The tender's hull was of Florida live oak and white oak, topped by a flush deck fore and aft. She measured 140 feet overall, 22 feet in beam and drew 9 feet. When loaded the *Shubrick* displaced 305 tons.[25] The tender's skin was painted black, topped with a white ribbon and waist. Red paddle wheels, white paddle boxes, and a black bowsprit, yards, and gaffs added "a saucy touch to her long and graceful cutwater," with six inches of bright copper shining above the waterline.[26]

A single expansion steam engine drove the *Shubrick*. Three furnaces heated a boiler measuring twelve feet long, eleven feet in diameter, and providing steam to a cylinder fifty inches in diameter. The tender was rated at 284 horsepower, and her churning paddle wheels could move the *Shubrick* through the water at eight knots. One former skipper claimed, given the right conditions, she could, when pushed, make twelve knots.[27]

The *Shubrick* was rigged as a brigantine, with her forward mast fully rigged and her mizzen gaffed. She also carried a foretopgallant sail and flying jib. Unlike other tenders, the *Shubrick* had armament consisting of one twenty-four-pound Dahlgren cannon mounted on a pivot carriage on the forecastle; and two Dahlgren twelve-pounders from the stern ports.[28] This unusual addition to the tender was "in case of incursions of the Indians from the British dominions in the Straits of Fuca and vicinity, to protect the keepers and citizens in that quarter against their attacks."[29]

If the attackers should manage to make it to the decks of the *Shubrick,* then she was "supplied with a very novel and ingenious apparatus for throwing scalding water. . . . The apparatus is hose-shaped and made of copper, disjointed at numerous places, and connected with spiral springs, which are lined with brass, thus allowing it to be turned in any direction. There is an external covering of leather. The water is, of course, from the boiler. . . ."[30]

On 23 December 1857, dockside workers in Philadelphia cast off the *Shubrick*'s lines, and she began to make her long way around the Horn to San Francisco. Capt. T. A. Harris, the first skipper of the new tender, commanded nine officers and a crew of twenty-six. After a long and event-filled passage, including an attack of yellow fever in which an assistant engineer died, the *Shubrick* arrived in San Francisco on 27 May 1858, 155 days after departing Philadelphia.[31]

The next three years saw the *Shubrick* constantly under way, establishing new buoys in the waters of the West Coast and carrying supplies for the building of new lighthouses. In June of 1859, the tender became the first and only steamer of considerable size to navigate the mighty Columbia River almost to its headwaters.[32] By 1860, the U.S. Lighthouse Board felt there were enough buoys established so that the *Shubrick* was "laid up a greater part of the time for want of funds."[33] The tender was then recommissioned late in 1860 to rework the buoys, and according to the U.S. Lighthouse Board, the busy *Shubrick,* "in addition to her regular duties," had "done good service in affording protection against the Indians at many points along the coast."[34]

The 146-foot Holly, launched in 1881, and a small steam launch pull up to a light station. Light keepers and their families generally looked forward to the arrival of the tenders, with one exception: they also delivered the district inspector for his quarterly inspections.

In a bizarre twist, the cannon aboard the *Shubrick* actually provoked a greater threat to whites than Native Americans. On 30 July 1861, Victor Smith arrived at Port Townsend, Washington Territory, as collector of customs. Smith is described as the local "Judas Iscariot," and a man who "caused more turmoil, dissension and hatred than any other figure in Pacific Northwest History." Smith is the archetypical nineteenth-century developer. The new collector disliked Port Townsend and wished to develop the town of Port Angeles further to the west. The *Shubrick* entered the picture when Smith traveled to Washington, D.C., in May 1862. Smith placed Lt. James H. Merryman, of the U.S. Revenue Marine, in charge as deputy collector of customs. In Smith's absence, Lieutenant Merryman accused Smith of mishandling funds, or "defalcations" as he termed it. In the meanwhile, Smith, in a neat little maneuver while in the nation's capital, managed to have the Customs House transferred to Port Angeles. This, along with shortage of funds, caused a stir—to say the least—in Port Townsend. When Smith left the *Shubrick* in Port Townsend to retrieve the records, Lieutenant Merryman refused to turn over the material. "In a huff," Smith returned to the *Shubrick* and ordered her cannon trained on Port Townsend, "threatening to bombard" the settlement "unless the records were given to him at once." Merryman surrendered. Smith retrieved the records and sailed to Port Angeles.[35]

The *Shubrick* served during the Civil War in the U.S. Revenue Marine. In 1865, the tender found herself transferred to the navy and designated the flag ship of an expedition made up of five other ships. The small fleet's orders sent it northward to make the initial survey for an attempt to lay a telegraph cable across Bering Strait. The attempt to lay the cable was aborted as the completion of the Atlantic cable made the Bering Strait connection unnecessary. Shortly after that, the navy returned the former tender to the U.S. Revenue Marine and on 24 December 1866 the secretary of the treasury ordered the *Shubrick* returned to the U.S. Lighthouse Board.[36]

While transporting construction supplies to the new light being built at Cape Mendocino, the *Shubrick* went aground in heavy fog about thirty miles below the cape on 8 September 1867. The captain considered her a total loss but not so T. J. Winship, chief engineer. Winship was a "plank owner" of the *Shubrick* and pleaded with the skipper to let him attempt to salvage the ship. The engineer gained permission and, in a remarkable feat, "took out all the machinery, cut out the boiler and discharged the guns, stores and supplies. He then raised the vessel up, placed her on 'skids' and transported her about 500 yards along the beach to a point for launching. All the machinery, etc., was placed on board and vessel successfully launched."[37]

The U.S. Lighthouse Board, meanwhile, felt the *Shubrick* a total loss. The board requested and received appropriated funds for a new tender. Unknown to the U.S. Lighthouse Board, *Shubrick* sailed into the Mare Island Navy Yard and was rebuilt almost completely for $162,399, nearly three times the original cost of the tender.

Protesting strongly to the navy over the cost, the U.S. Lighthouse Board nevertheless paid up with the appropriated funds for the new tender.[38]

The *Shubrick* returned to working buoys again. The number of buoys grew so quickly that it was nearly impossible for one tender to adequately service them. In 1869, the U.S. Lighthouse Board began to petition Congress for another tender for the West Coast. Eleven years later, the *Manzanita* steamed into San Francisco and began the duties of a Twelfth Lighthouse District tender. The *Shubrick* then moved to the Thirteenth District, encompassing Oregon and Washington, with her new home port in Portland, Oregon.

By 1880, the *Shubrick* had begun showing her age. Six years later in 1886, the steam-powered screw tender *Madrona* arrived in San Francisco. The U.S. Lighthouse Board ordered the new tender to take up duties in the Twelfth District, while the *Manzanita* sailed northward to relieve the *Shubrick*. The *Shubrick* was sold on 20 March 1886 for $3,200. The new owner towed the former tender to San Francisco, beached her on a sandbar, and removed anything of value. Then he burned the hulk to salvage the copper and brass that had gone into her construction. Thus ended the career of the first lighthouse tender on the West Coast.[39]

Like the aids to navigation that they serviced, the lighthouse tenders suffered during the Civil War. The records of the service show at least nine were captured by Confederate forces. One, the *Knight*, was seized in 1861 and retaken later in the war.[40]

THE U.S. LIGHTHOUSE SERVICE, as late as 1917, established four classes of tenders based upon length and draft. Shortly after that, the system changed, and the method used was by area of operation: coastwise, bays and sounds, river tenders, and buoy boats. Only an overview of the early coastwise class follows.[41]

Between 1878 and 1903, most of the ships in the coastwise tender class tended to be from 140 to 200 feet in length and possessed from 250 to 900 horsepower, with both single and twin screws and hoisting capacities of from eight to twenty tons.[42] From 1900 to 1908, five tenders were built that had certain features in common. One such feature was buoy ports at the midpoint of the buoy deck, which permitted swinging buoys aboard with the minimum lift over the side and pads of steel plate and angle construction, with wood filling to provide protection for the ship's side. Rubbing strakes ran from stem to stern at the waterline to protect the ship. Finally, the pilothouse was built to give maximum visibility and each of the vessels were fitted with a superintendent's quarters aft.[43]

In 1908, a class of eight tenders came down the ways with the same specifications, known as the *Manzanita* (there were three tenders with this name), or eight tender class. The major change in these vessels was the boom and "hoister mechanism since steel was replacing wood and wire rope replacing manila [line]." This class also had a low and "turtle back" forecastle, which gave the ship handler a better view when trying to get alongside a buoy.[44]

At 207 feet, the Cedar, built in 1917, was the largest lighthouse tender ever built by the U.S. Lighthouse Service. The tender's operating area was Alaska.

Generally, the tenders up to 1917 had short operating ranges. For operations in Alaska, the *Cedar* was built. At 201 feet in length, with a 36-foot beam and displacing 1,970 tons, she was the largest tender built by the Lighthouse Service. The *Cedar* had a single screw and had 1,150 horsepower and drew 14 feet. The low forecastle was not a part of the design, but one innovation was housing the anchor within the hull to prevent it from hooking on a buoy when coming alongside.

In 1919, Commissioner George R. Putnam was forced to accept army mine layers for conversion to buoy tenders. The turtle back deck was installed, and the anchor was mounted high to prevent damage to buoys; all hoisting machinery was installed below decks to prevent deterioration.[45]

Between 1930 and 1939, the *Violet, Arbutus,* and *Hollyhock* classes were the last classes of tenders built by the U.S. Lighthouse Service, with *Juniper* being the last ship built. The latter vessel was the first all-welded-steel and diesel-electric-propelled coastwise tender.[46]

After taking over the U.S. Lighthouse Service, the U.S. Coast Guard continued the color scheme established by the service for the tenders—black hull with a white superstructure. It was not until 1947 that "lighthouse tenders" were called "buoy tenders."[47]

THE SETTING OF A LARGE SEA BUOY by a buoy tender's crew is a demanding and potentially dangerous procedure. To set a large nine-foot-by-thirty-two-foot sea buoy involves the following steps. The buoy is usually lying horizontally on deck and made ready for lighting. An anchor chain is fastened to a bridle under the buoy, and the chain is faked out on deck. The heavy sinker is suspended at the forward end of the buoy port. The buoy is then hoisted nearly horizontally with two purchases by the boom and swung out to clear the side of the buoy tender. The lower purchase is slacked off until the hook is holding the buoy nearly vertically. A line is then passed through one of the lifting lugs on the buoy to hold it close to the ship when lowered into the water. Once in the water, the lens is installed, and the light lit. The buoy tender is maneuvered to the correct position, and the sinker cut loose. Fathoms of chain rush across the buoy deck; there is the deafening sound of metal against metal, and flecks of scale fly through the air raised by the chain's plunge into the depths. With modern booms and hoisting gear, the above may sound fairly straightforward, but when combined with a rolling or pitching deck, the chance for danger enters the picture.

The main mission of the tenders during the U.S. Lighthouse Service years was to service lighthouses and buoys. Much like the lightkeepers on land and on lightships helped those in distress close to their units, however, the sailors aboard the tenders also performed errands of mercy. In January 1916, the tender *Columbine* intercepted a distress call from the bark *British Yeoman* just off Port Allen, Kanai Island, Territory of Hawaii.[48] The *Columbine* managed to locate the bark during the night and found her with no anchors, the rudder carried away, and her stern approaching the breakers near the beach.

The *Columbine*'s skipper, Capt. Frank T. Warriner, personally took charge of the tender's whaleboat and brought the boat and its crew into the "boiling breakers four times" in an attempt to pass a heavy towing hawser to the stricken bark. Each time a strain was put on the hawser, it snapped.

The tender, much older and smaller than the *British Yeoman,* could not tow the bark. (The *Columbine* at the time was twenty-four years old.) Strong winds and seas continued to lash the two craft. Captain Warriner radioed for assistance the next morning and was informed the navy tug *Navajo* was en route. Because of the seamanship of Captain Warriner, the *Columbine* managed to haul the bark clear of the beach. The *Navajo* arrived on scene and took over the tow, with the *Columbine* escorting both ships to Honolulu.

In a letter of commendation, Secretary of Commerce William C. Redfield wrote, "Despite darkness and storm, undismayed by the heavy seas or by the repeated breaking of hawsers, the courageous crew of the tender stood steady at their tasks for 56 hours without let-up until the bark was safe. I bring this incident to the attention of the entire Lighthouse Service to make it an example to all of unselfish devotion to duty."

One example of heroism by an individual aboard a tender occurred on 17 September 1877, as the lighthouse tender *Rose* lay in the Christiana River at

The spruced-up crew of the Cedar *pose beside buoy sinkers (in the right foreground). Such photographs of lighthouse tender crews are rare.*

Wilmington, Delaware. Capt. Charles H. Smith happened on deck and glanced shoreward toward the wharves. His gaze fell upon a young boy in the river who was "on the point of drowning in the presence of a number of bystanders."[49]

Without "taking off coat, hat, or boots," Captain Smith "tumbled" over the side of the *Rose* and struck out for the hapless boy, who was "some distance" away. Captain Smith grabbed the child, worked toward the nearest wharf, and handed him up to the bystanders. The skipper of the *Rose* then "nonchalantly" swam back to the tender and managed to clamber up the side. Once on deck, he went below to remove his wet clothing. Apparently Captain Smith's attitude of having "done nothing to speak of" greatly impressed witnesses. It also impressed others, for Capt. Charles H. Smith won the Silver Life Saving Medal, the Treasury Department's second highest award for saving people from drowning.

The reality of work on board a tender belies the neat appearances of crewmen in posed photographs. Here the crew of the Tulip *transfers coal to another tender, which will probably deliver the cargo to a light station. The clothing of the men and their expressions graphically show that coaling was exhausting, dirty work.*

IT IS IMPORTANT TO NOTE how valuable the ungraceful tenders were to keepers of lighthouses and their families, especially to those at isolated locations. While the tender may have brought the stress of the district superintendent and his inspection, it more importantly brought pay, mail, medical help, supplies, and relieving keepers. More than one former resident of a light station has remarked upon the excitement generated on the visit of the tender.[50]

In 1939, the U.S. Lighthouse Service had a preliminary design for a new, versatile tender design. The 180-footer, in addition to her regular requirements, had an open-sea search-and-rescue capability. When the U.S. Coast Guard took over, icebreaking features were added to the design. The 180s, known as the *Cactus,* or 180(A), class, were built at either the Zenith Dredge Company or the Marine Iron and Shipbuilding Corporation, both located in Duluth, Minnesota. The class has a beam of thirty-seven feet and a draft of thirteen feet. Full load displacement is 1,025 tons. The tenders are diesel-electric and produce 1,000-brake horsepower on a single screw for a maximum speed of 13 knots. The boom has a twenty-ton capacity, and the buoy deck is capable of sustaining loads of up to fifty-four tons. All the *Cactus*-class keels were laid in 1941–42 and placed into commission during 1942–43.

In 1990, nine of the thirteen buoy tenders in this class were still in commission.[51]

The tenders entered the maritime world during World War II. No one could ever say the tenders were the U.S. Coast Guard's offensive threat to the Axis powers, but they did prove to be valuable in the war effort. For their combat role, tenders received a gray paint scheme and were armed with one three-inch gun aft, depth charges, and machine guns. In the Atlantic, seaworthiness allowed the ships to perform convoy escort and, after the establishment of ocean weather stations, the 180-footers manned the stations, thus freeing larger U.S. Coast Guard cutters to perform convoy duty. (Ships established in various locations in the North Atlantic transmitting weather observations became known as ocean weather station ships. The observations supplied weather forecasters better information upon which they could base their predictions for military operations.) During World War II, the tenders played a major role in the Greenland area. The ships, with ice-reinforced hulls, helped to escort other ships through the ice-choked waters. The war did not stop the tenders from the normal duties of caring for buoys. If anything, the amount of shipping needed to sustain a global war effort increased the demands for buoys.

In the Pacific, the new class of tenders proved how versatile the ships could be. The ships served as supply vessels, dispatch vessels, construction tenders, net layers, fire fighters, salvage vessels, rescue vessels, and, of course, as buoy tenders. The Pacific at the time of World War II was still poorly charted and in need of aids to navigation. The *Balsam,* for example, logged fifty equator crossings in nineteen months with duties ranging from establishing aids to navigation and building loran stations to general supply work.[52]

In September 1943, the *Buttonwood* worked aids to navigation at Guadalcanal and then sailed to Australia. There she also worked aids to navigation, with some time out to rescue the SS *Minjak Tanah* grounded on the Great Barrier Reef. In 1944, the *Buttonwood* sailed for New Guinea to establish buoys and then to the Phillippines with supplies. While there, she came under several aerial attacks. During one air raid, her gunners claimed hits on a Japanese bomber, and four of the tender's crew received injuries from shrapnel. The next day, three more crewmen found themselves in sick bay after another raid. A month later, *Buttonwood*'s gunners downed a Japanese fighter. That same afternoon, the tender ran aground on an uncharted coral reef. The damage-control party quickly went to work and shored up the ship, and she resumed normal operations.

On Christmas Day 1944, the *Buttonwood* went to the aid of the SS *Someldijk,* ablaze after being hit by an aerial torpedo. The tender's skipper, Lt. R. W. Fish, carefully laid his ship alongside the *Someldijk* and then put fire-fighting parties aboard the burning vessel. One hold was ablaze with lumber and the adjacent hold contained ammunition. The "between bulkhead heated a cherry red," but the tender's crew battled the fire throughout the night before finally bringing the inferno under control. Adding to the danger, the area was under an air-raid alert.[53]

The tenders faced danger from the air and from beneath the sea. The *Acacia* was sunk in the Caribbean by the German submarine U-161 on 15 March 1942. The *Conifer* en route to Boston sighted a periscope on 8 August 1943, commenced two attacks, and then lost contact.[54]

The *Papaw,* under way from Norfolk to the Canal Zone and then into the Pacific, made a sonar contact on 21 April 1944. The tender attacked. After her run, the *Papaw* returned to the contact, and the crew spotted a large area of bubbles. Then a "dark cylindrical object, about 60-feet long," surfaced about 500 yards off the tender but sank before the *Papaw*'s guns could open fire. Sonar made no further strong contact, "although discolored water and oil showed on the surface." The *Papaw* then changed course to Guantánamo Bay, Cuba, to replenish her depth charges and continued on to the Pacific.[55]

In the Pacific, the *Papaw* had some unusual experiences. For several months, the tender removed wrecks in the Saipan area. One night, her small boat crew approached a wreck, only to find two Japanese soldiers hiding there. The boat crew opened fire, killing one soldier while the other escaped in the darkness. En route to Peleliu Island in October 1944, the tender survived a typhoon, although her deck cargo broke loose in the tempest. At Palau, on Christmas Eve 1944, the *Papaw* suffered a four-foot-long and two-foot-wide hole in her fore peak near the water line from a Japanese mine. The tender made temporary repairs and continued to work aids to navigation.[56]

The *Ironwood*'s experience with submarines was different from that of most ships during World War II. The tender sailed to Cape Esperance on 3 January 1945 and embarked two navy divers and material needed to salvage a two-man Japanese sub-

The 180-foot tender Ironwood *helped raise and tow a Japanese two-man submarine near Guadalcanal during World War II.*

Buoy tenders continue their traditional duties; a buoy is brought aboard the tender Redwood.

marine sunk in thirty feet of water. After three weeks of work, the *Ironwood* raised the midget submarine, placed it alongside, and moved to a navy crane barge to hoist the sub aboard the barge.[57]

With the end of World War II, the buoy tenders took up their normal peacetime duties of servicing aids to navigation. Conflict in the Pacific area again brought the tenders into a combat region. During the Vietnam War, two buoy tenders serviced aids to navigation along South Vietnam's coast, rivers, and harbors.

In 1995, the U.S. Coast Guard began adding new buoy tenders to their fleet. A new "Keeper" class, named the *Ida Lewis,* was launched on 14 October 1995. The tender is 175 feet in length, with a beam of 36 feet. The tender's forty-two-foot boom can lift ten tons. The missions listed for the tender are: aids to navigation, marine environmental protection, search and rescue, and domestic ice breaking. The cutter is 225 feet in length, 46 feet at the beam, and has a draft of thirteen feet. The buoy tender is named for Ida Lewis, the long-time keeper at Lime Rock, Rhode Island (see chapter 4). The *Juniper,* launched in July 1995, marks the first of a new class of seagoing tenders.

By 1996, U.S. Coast Guard tenders were divided into seven classes based upon size and tending capacity. The first class of tenders, the seagoing, are 180 feet in length and capable of lifting up to twenty tons. They are equipped for long voyages and have ice-breaking capability. Coastal tenders make up the next group of

Work on buoy tenders today remains difficult, dirty, and dangerous, as shown on the buoy deck of the Redwood. *A cement sinker is in the foreground; part of the buoy is at the right edge of the photograph.*

vessels and are 133, 157, and 175 feet in length and can lift ten tons. Inland tenders are divided into two classes, large (100 to 131 feet) and small (65 to 91 feet). The large tenders can lift ten tons. River tenders are also divided into two classes: large (104 to 115 feet) and small (65 to 75 feet) and can lift ten tons.[58]

The hardworking, unglamorous, black-hulled buoy tender is usually overlooked by those who write and photograph the sleek warships of the U.S. Navy or commercial ships, as mentioned previously. Yet, most ships would have difficulty reaching port safely without the necessary work on aids to navigation carried out by these vessels. The tenders have faithfully serviced aids to navigation, whether under the U.S. Lighthouse Service or U.S. Coast Guard. In time of war, the crews of the black fleet performed as well as sailors aboard the larger and much heavier armed warships and still carried out their mission of ensuring that shipping had the aids to navigation necessary to carry out their mission. One historian of aids to navigation wrote: "In the Pacific [during World War II], it was not uncommon for a tender's crew to be working a buoy off the starboard side, while their shipmates fought off Japanese *Kamikazes* to port."[59]

Fog Signals and Fancy Buoys

THE PICTURE MOST PEOPLE HAVE of lighthouses is a light piercing the darkness; they imagine the deep growl of a fog horn echoing through the mist. Unfortunately for romantics, this is not a completely accurate picture. Many light stations did not have fog signals. On the East Coast, for example, few fog signals existed south of the Chesapeake; however, most of the lights north of the Chesapeake Bay did have fog signals. Simply put, fog signals were placed where there was a probability of encountering the murk. George R. Putnam, for example, reported the light station at Seguin Island, Maine, logged 2,734 hours of fog in one year—one out of every three hours were fog shrouded at the station. At locations where fog was not a factor, there was no need to install a signal.[1]

The first fog signal in what would become the United States was at Boston—also the location of the first light station in this country. The fog signaling device was a cannon that went into service in 1719. The use of cannons to help ships find their way through fog is not unusual. England still tested cannons as fog signaling devices as late as the 1860s.[2]

By 1857, the U.S. Lighthouse Board hired a retired army sergeant to fire the first fog signal on the West Coast at Point Bonita, on the opposite side of San Francisco's Golden Gate. San Francisco's heavy fogs caused the harried ex-sergeant to get only two hours rest during one period of three days and nights. During the first year the sergeant fired 1,390 rounds, expending 5,560 pounds of black powder at a cost of $1,487. After the weary sergeant pleaded for help, the district superintendent sent for an assistant, and between the two men they used up $2,000 worth of black powder, three times the sergeant's salary. The always cost-conscious U.S. Lighthouse Board discontinued the signal. According to the board, the use of a cannon as a fog signal stopped in this country "because of the danger attending its use, the length of interval between successive explosions, and the brief duration of the sound, which renders it difficult to determine its direction with accuracy."[3]

Bell boats also helped warn mariners trying to enter San Francisco's foggy harbor. Bell boats were in use in Delaware Bay in 1885. The boats, thirty feet long and twelve feet wide, were turtle backed. Suspended under deck scaffolds, the bells weighed a half-ton. Four clappers struck the bell as the boat moved with the waves. The boats were crewless and were held in place by heavy moorings. By 1887, the craft were no longer in use.[4]

In the 1820s, fog bells came into use at many New England lighthouses. The bells at first were struck by hand, a task that must have been unpleasant to the ears. Around the 1860s, the U.S. Lighthouse Board installed devices that rang them mechanically. The board found that a clockwork system—using a falling weight as a source of power—to be the most practicable. The early devices, however, were noted for constantly breaking down.[5]

In 1906, Juliet Nichols, the keeper of the Angel Island light, located in the San Francisco Bay area, observed the normal fog rolling in through the Golden Gate. She started her mechanical fog bell only to have it fail. Juliet could "see the masts of a sailing vessel approaching above the fog."[6] She then grabbed a hammer and started to hit the bell at the prescribed intervals. Twenty-four ear-numbing hours later the fog lifted, and Juliet could rest. Two days later, the mechanical device broke again, and Nichols had to repeat her extraordinary work. The bell, as of 1990, was the only artifact left of the original light station. It bears the marks of Juliet Nichols's hammer.[7]

The U.S. Lighthouse Board over the years sought to perfect the devices that rang the fog bell. By the 1920s, an electrically operated bell was in service, as was a device that automatically turned on the bell in thick weather. The device, operated by a hydroscope, used hair to measure moisture. When the humidity approached 100 percent, as it does during fog, snow, or rain, the hydroscope would trip a switch, which turned on the electrical current to the power for the bell. To the embarrassment of the U.S. Lighthouse Service, the device displayed its use in Baltimore's harbor in the 1920s. President Warren G. Harding was to dedicate the Francis Scott Key Memorial at Fort McHenry. In honor of the occasion, fire boats shot streams of water in the air. At the nearby U.S. Lighthouse Service depot at Lazaretto, an automatic fog bell stood guard. The wind carried the moisture from the spray and activated the hydroscope. As the speakers began their speeches, the two thousand-pound bell began to sound off.[8]

The U.S. Lighthouse Board experimented with several sound devices designed to warn the mariner in fog. Locomotive whistles shrilled, but the sound apparently was "diffused" too quickly. Celadon L. Daboll of New London, Connecticut, developed the first practical power-operated fog signal. The compressed air signal could be operated by, strangely enough, a horse or by hand. The horse could either walk up a ramp that compressed air into a holding tank, or the animal could walk in a circle

Charles Nordhoff drew this sectional view and called it "Operation of a Siren."

to compress the air. An early drawing of the device at the Beavertail Light Station depicts a horse in a large wheel-like device that can be seen on a smaller scale in pet hamster cages. Compressed air could also be pumped into tanks by hand. The sound was generated by the compressed air that vibrated a reed ten inches long and more than two-and-a-half inches wide.[9]

Experiments with steam-powered whistles began in 1855. By 1867, the U.S. Lighthouse Board reported the need for powerful steam signals, but hesitated to install many of the signals because of the cost, as always, and also because of "the danger of intrusting the management of an agent of so much explosive power to ordinary light-house keepers." The first light stations regularly equipped with the steam whistles were West Quoddy Head and Cape Elizabeth, both in Maine. The signals employed a ten-inch locomotive or ship whistle.[10]

Another experiment utilized sirens. First installed at Sandy Hook East Beacon in 1868, the device had a large cast iron trumpet and its mouthpiece consisted of a

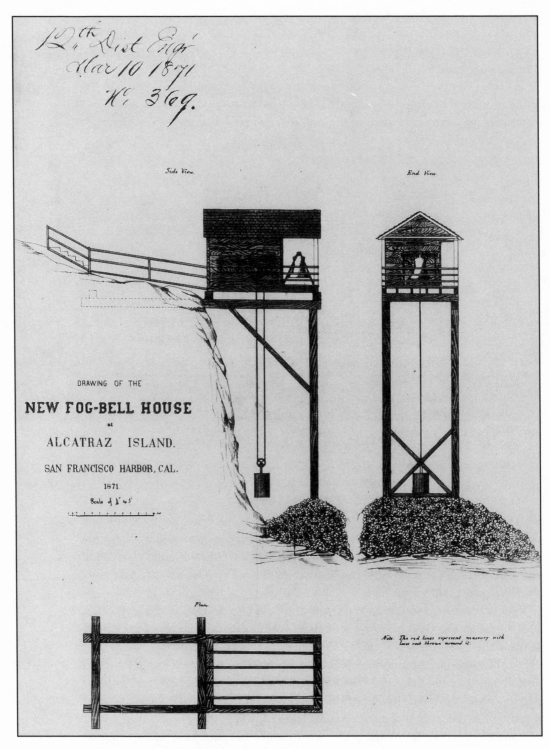

This is an 1871 drawing of the fog bell used at Alcatraz.

fixed, slotted disc. Another revolving disc was placed behind the fixed disc. The chamber to which the trumpet was attached was the dome of the boiler. Approximately seventy pounds of steam were forced through the discs to produce the distinctive sound. A reporter, upon first hearing one of these sirens, described it as having "a screech like an army of panthers, weird and prolonged, gradually lowering in note until after half a minute it becomes the roar of a thousand bulls, with intermediate voices suggestive of the wail of a lost soul, the moan of a bottomless pit and the groan of a disabled elevator."[11]

Experiments also utilized a natural orifice. On the Farallon Islands, some forty miles off the harbor of San Francisco, Maj. Hartman Bache, the West Coast district inspector, built a brick structure in 1859 near the edge of a blowhole. The water rushed out of the blowhole and rushed through the structure, pushing air before it. The air then sounded a locomotive whistle.[12]

By the 1870s, the siren trumpet, bell, and whistle were the accepted fog signals in the United States, with the bell used in such inland waters as the Chesapeake Bay. By 1900, there were 377 fog signals in the United States, exclusive of those on buoys. By the turn of the century, the electric siren was developed to replace the steam device. The diesel engine became another new invention to help power air compressors for fog signals. Both of these innovations helped relieve the keepers of the back-breaking drudgery of handling coal to power the boilers for steam; the signals could start much faster than the ten minutes or more required to get up steam for the signal.[13]

Around the turn of the twentieth century, a Canadian company developed the diaphone signal in three models: single tone, two tone, and chime. The U.S. Lighthouse Service was so taken with this signal when it was introduced into the United States in 1914 that it acquired the rights to manufacture the equipment in this country. The work was undertaken in New Jersey. The deep sounds emitting from these signals were more "population friendly," and district offices received fewer complaints from the sleepless populations of maritime cities plagued by fog. The diaphones, however, were expensive and required a technical skill for repair that many keepers did not possess. In 1929, the service developed a diaphragm signal. This horn sounded almost like a diaphone, but was less expensive to build and repair.[14]

The U.S. Coast Guard began to phase out many of the older devices after World War II, replacing them with new electronic horns. The greatest change, however, came about when the International Association of Lighthouse Authorities decreed "that fog signals are no longer necessary for the needs of navigation," so now "a few diaphragms and the new electronic pure tone signals" remain in some areas where "fishing fleets or pilot pressures are strong." The fog signals have gone the way of manned light stations.[15]

Next to lighthouses, probably the most recognizable aid to navigation is the buoy. The landsman may not know that the old navigator's ditty, "red, right, returning," means to keep the red buoy on the starboard side when returning from sea, but he can recognize a buoy. Buoys not only mark hazards and channels, but their size, shape, and color scheme impart important information to the mariner.

The first use of buoys in northern Europe were in the Vlie River, which empties into the Zuider Zee. These early fifteenth-century markers guided ships into Amsterdam and Kampen. In sixteenth-century Sweden, the Pilot Service Organization was responsible for maintaining markers on shore and on shoals. It is not clear as to whether the shoal-water markers were buoys or structures, but at least some buoys were used. Some markers were hollow, wooden casks bound with iron bands and moored with chain and a large stone. Others were regular brooms or the shape of brooms, which apparently was the symbol for an aid to navigation of the time.[16]

Several buoys marked hazards in Delaware Bay as early as 1767. There is a notation of four buoys transferred from Massachusetts and three from Pennsylvania to the new federal government in 1789. The buoys were either small barrels or bundles of wood bound with rope or an iron band.[17]

By 1808, at least seventy-seven buoys aided mariners. The devices were made of wooden staves about four feet in diameter and five feet in length and held together by iron rings attached to a mooring chain. As late as 1871, a British lighthouse report commented upon this type of aid to navigation: "Some buoys . . . disappear under water as soon as the tide becomes strong and only reappear at slack water. Generally speaking, the buoys in use are not constructed on scientific principles; there are others either used or designed which show more thought."[18]

The first wood spar buoys (long, telephone pole–shaped devices) began to see service in the United States in 1820 and were weighted at one end to make them stand upright in the water. Wood continued to be the mainstay for buoys until the middle of the nineteenth century, when the boiler iron came into use. New iron buoys were compartmented to make sinking difficult. The first bell buoy came into service in 1855. Wave action activated the bell clappers, basically on the same principle as the bell boats. This same year saw at least 1,034 buoys in U.S. waterways. By 1876, the first whistle buoy was on station, and five years later a combination oil-and-gas-lighted buoy commenced operations.

Testing of electric buoys began between 1888 and 1903, with power supplied by running a cable underwater to a row of wooden spar buoys. This experiment was not a success. Batteries would eventually replace any need for a cable. By 1900, the number of buoys in the United States had risen to 4,842. The first tall-type nun (conical-shaped) and can (flat-topped) metal buoys came into use the same year. Acetylene gas power buoys came into use in U.S. waters ten years later. By 1939, there were approximately 16,000 buoys in U.S. waters.[19]

Buoys are held in place with chains attached to a bridle using shackles and swivels. The chain originally fastened to a stone. Later standard ship's anchors replaced stones, which were replaced by mushroom anchors. The device now used to hold a buoy in place is a large concrete block known as a sinker. Buoys are pulled at least once a year to check the chain and attachments, plus to give the buoy itself an over-haul. In areas of freezing, buoys must be removed each year before freeze-up and then replaced in the spring.[20]

The landsman may notice that buoys, besides coming in different shapes, also are of various colors. The establishment of a color scheme for buoys came from England in the middle of the nineteenth century. The color of the buoy speaks a language of its own for the navigator. When returning from sea, red marks the right-hand side of a channel and green (formerly black) designates the left. Red-and-green

The U.S. Lighthouse Service maintained bases in each district to keep extra buoys available for emer-gencies. This early photograph of the Staten Island, New York, depot shows tenders, a lightship, and, on the dock, nun buoys (conical shaped), can buoys (the flat objects marked #3, next to the striped buoy), and a center-channel buoy (white striped), along with a mooring chain and sinkers. On the dock next to the lightship are a Courtenay's whistle buoy and two riveted iron spar buoys.

horizontal bands (again, formerly black) mark junctions in the channel. If the top band is green, the preferred channel "will be followed by keeping the buoy on the port hand" and vice versa if the top band is red.[21]

Other colors also help the navigator. White buoys mark anchorage areas, while white buoys with green tops are used to designate dredging or survey operations. White-and-black horizontally banded buoys mark fish net areas.[22]

Along with a color scheme, buoys are numbered also to indicate which side of the channel they mark and to help the navigator find the buoy's charted location. Numerals increase from seaward and are kept in sequence, with even numbers on the right-hand side of the channel and odd on the left.[23]

Buoys have always run afoul of ships. At one time tugs towing barges were constantly passing close aboard buoys and fouling their manila towing hawsers in the buoys, which would then either be damaged, pulled off station, or run over. The constant repair or replacement of these expensive aids to navigation concerned the U.S. Lighthouse Service. The service had metal saw teeth placed on the buoys that were most likely to be damaged by the tow lines. Tug operators soon found that their expensive hawser could be ruined and most began passing the buoys with more leeway.[24]

Whether by an act of nature, a break in a chain, or being hit by a ship, buoys have gone adrift. Some have had remarkable travels. Nun buoy Number 1, moored at the Columbia River bar went adrift in a gale in January 1889. It was picked up in June nearly two thousand miles away at the mouth of the Karluk River in Alaska. A buoy broke its moorings off San Francisco and seventeen months later was retrieved off the coast of Maui, Hawaii. Another buoy at Matanilla Shoal in the Florida Straits drifted off station during a hurricane in 1926. The buoy, with its mooring still attached, drifted for several months half submerged before a buoy tender retrieved it.[25]

The twentieth century ushered in an age of electronics that has had a huge impact upon aids to navigation. The first major device was the radio beacon. Radio beacons can guide the navigator through the thickest fog and can also help ships out of sight of land determine their position. Commissioner Putnam wrote that the radio beacon was one of the "most notable" advances and the most significant contribution to aids to navigation during his tenure.[26]

World War II hastened the development of electronics in position-finding and the system of LORAN (Long Range Aids to Navigation) was another step toward the reduction of light stations in the United States. For hundreds of years, a good chronometer and sextant remained the best navigational tools available to the mariner at sea. Normally, these instruments were satisfactory, but there were problems. The steps used to take celestial observations with a sextant and to place the position on a chart were laborious and time consuming. The time period in which stars can be used for fixing a position is narrow—usually in twilight—and this is further hampered by weather. Anyone who has gone to sea knows it is not uncommon for cloud cover to stop any observation of the sun or stars for several days.[27]

World War II sharpened the need for obtaining accurate navigational fixes, especially in the vast reaches of the Pacific Ocean. Electronics engineers of the U.S. Coast Guard and U.S. Navy began developing theories that would eventually lead to a chain of LORAN stations. The theory is based on the speed of light and radio waves—both travel at 186,00 miles per second.

A controlling station, designated the "master," transmits a pulsed signal that is radiated in all directions. The signal travels along a straight line until received by a secondary unit called the "slave" station. At the same time, the signal travels to the position of any ship or aircraft in the area, which may be equipped with a LORAN receiver. The receiver converts the signal to a visual display on an oscilloscope, which looks like a small television set. The slave station also receives the master signal, waits a specified amount of time, and then transmits a pulsed signal of its own. This signal is then picked up by the receiving station where it appears as a second trace on the oscilloscope. The LORAN receiver is arranged so that the master signal appears on top and the slave on the bottom and slightly to the right of the master. Rotating control knobs on the receiver set positions the slave signal directly below the master, and the knobs eventually cause the superimposing of the two signals. When this is completed, the navigator records a number from a direct-reading counter on the set. This number indicates how many microseconds happen between the time a master signal reaches a receiver site and the slave signal triggered by that master signal arrives at the same point. With this number, the navigator now refers to a specially printed LORAN chart of the general areas in which he is located. The chart is overprinted with numerous lines, each of which represents a LORAN time difference. The lines are labeled in microseconds. By matching the reading with the lines on the chart, the navigator determines his position is somewhere along one of the lines. Because the lines usually extend for long distances, a single line with no other information is almost useless. By taking readings on two or more master-slave combinations, however, the navigator can determine his position with accuracy.

A good navigator can usually obtain a fix within two to three minutes. Weather normally does not affect LORAN, which is usable at daytime ranges of up to 750 nautical miles and at night at almost double that distance. This system is known as LORAN A. The U.S. Coast Guard also developed a "C" system with accuracy to within a quarter of a mile and ranges above three thousand nautical miles.[28]

"There are," however, "deficiencies in basic LORAN operation." These deal with how radio waves travel—along the earth's surface and via the sky. A skywave takes longer to reach the set and can introduce a margin of error.[29]

The LORAN wave also travels over land and water, two entirely different mediums. LORAN assumes the speed of a radio wave to be constant, when in fact the speed is altered when a wave is over land. Lastly, LORAN coverage does not include all areas.[30]

Work first started on the LORAN station chain in 1942, with the first stations set up by the U.S. Navy in cooperation with the National Defense Research Council

The first atomic buoy, tested in Baltimore harbor in 1961, proved unsuccessful.

and eventually taken over completely by the U.S. Coast Guard. The North Atlantic chain came into being in mid-winter of 1942 and then began in the Aleutian chain in the same year. Eventually, the service spread throughout the Pacific and the world.[31]

One of the greatest improvements in navigation through electronics is the recent use of satellites and the Global Positioning System (GPS). A navigator has only to press a button to receive a read-out of the position without the tuning of LORAN. The position is accurate to within feet.

The rapid acceptance and use of electronic navigational tools have helped ease some of the burden of command belonging to the captain of a ship. Even with the plethora of devices available to the navigator, the wise skipper will still make use of the low technology of the buoy.

Eight Bells

L ESS THAN TWO AND A HALF YEARS after the U.S. Lighthouse Service merged with the U.S. Coast Guard in 1939, the first of many major changes took place with the lighthouses, lightships, and lighthouse tenders of the former service. The first change came about because of a world at war. In the long view, however, the important transformations came about because of changing technology. This chapter will deal only with the changes in lighthouses, as lightships, tenders, and buoys have already been discussed.

After 7 December 1941, security required that some sea coasts' lights be shut down. A total of nineteen major lighthouses turned off their lights, and ninety-one major lights reduced their power. At least 1,227 minor lights shut off the power to their lamps.[1] Some areas apparently did not receive word to extinguish their lighthouses. On 14 May 1942, Col. George C. Van Dusen, Military Intelligence in Miami, Florida, wrote that a chance remark by a skipper of a torpedoed ship caused him to investigate the lights of eastern Florida. Finding that 90 percent of all the sinkings on the east coast of Florida occurred at night, and 65 percent of all the sinkings occurred "within the arcs of light thrown by the East Coast Lighthouses," he recommended that "lighthouses be extinguished for the duration, or if that was not feasible, that the radius of the light be greatly cut down."[2] In areas of danger however, some types of aids to navigation for assisting the mariner still needed to be provided. For example, Alaskan lights shut down, but by 30 December 1941, Allied shipping through the Inside Passage—the waters of southeastern Alaska and British Columbia—dictated their relighting.[3]

Most districts planned for complete blackouts or dimouts, which led to a major problem. A complete blackout required that navigational aids shut down, but the amount of work required to manually turn off the aids seemed daunting. By 1942, U.S. Coast Guard headquarters developed a system using a radio signal that was intended to black out unattended lights. A control station would transmit a special

In South Carolina, the Charleston Light Station (photographed in 1876, above, and again in 1958) vividly illustrates the fate of some light stations over time.

coded, ultra high frequency signal with a receiver mounted on the buoy or unattended light. This system, eventually becoming known as ANRAC, also helped signal light keepers to darken their lights. According to the U.S. Coast Guard's official history of the war, however, the system proved unsatisfactory for unspecified reasons in the Thirteenth District (Washington-Oregon).[4]

Light towers made logical lookout locations in which to spot possible offshore enemy craft, or actions, and many of the major stations received additional personnel to scan the immediate areas. In fact, lookout duties in the Third District (headquartered in New York City) began two days prior to the attack on Pearl Harbor. Crews assigned to thirty stations in the Third District had orders to report unknown warships, raiders, mines, or any suspicious activities or craft.[5]

On 1 April 1946, a tsunami (sometimes mistakenly called a tidal wave) swept over the Scotch Cap Light Station, which was ninety feet above the sea, completely destroying it and killing the five U.S. Coast Guardsmen at the station.

At Montauk Point, New York, a lookout spotted a plane that crashed into the sea and quickly relayed the message to a nearby coastal picket boat station, whose standby boat rescued the downed pilot. In the same district, the lookout at the Coney Island Light Station found five mines washed upon the beach of the resort area. The station's crew guarded the mines to prevent civilians from injuring themselves. The mines were removed and exploded at a safe location.[6]

The Thirteenth District believed that men serving at lighthouses needed military training as much as any other person serving in the armed forces. A mobile chemical warfare unit traveled to various light stations to give instructions in gas mask usage. In addition, the men received training in rifle marksmanship. At light stations near other military locations in the Thirteenth District, there were attempts to camouflage the light structures.[7]

In August 1944, as the war moved toward an Allied victory, the sea coast lights began to resume their normal operations, at least on the East and Gulf Coasts. When the war ended and each district reported its effects upon aids to navigation, personnel received the most negative impact. The war brought constant rotation of military men to the light stations, and "a morale problem was created when men, who loved lighthouse work, found themselves transferred to beach patrol, port security, or combat assignments, and when boatswain's mates who wanted to be at sea, or fighting the enemy far afield, were transferred to lighthouses on quiet shores."[8]

Automatic lights added to the growing move against manned light stations. As early as 1947, the U.S. Coast Guard opened the automatic Long Beach Harbor light. It had a 36-inch airway beacon-type lantern, with 149,000 candle power, a fog signal, and a radio beacon, but no crew. It became known as "Robot Light" and would be the future. The combinations of better navigational devices and automation were the real death blows to lights, but hastening it along was a budget-conscious U.S. Coast Guard. Facing chronic fiscal problems and low personnel levels, the service sought a cost-effective means of continuing to provide good aids to navigation for the mariner. With better electronic navigational devices, many lights were eliminated. The lights remaining could, in the view of the U.S. Coast Guard, be more effectively served by automation.[9]

The process of automation began gradually. By 1968, the U.S. Coast Guard introduced its Lighthouse Automation and Modernization Project (LAMP) to speed up the process. In 1995, twenty-seven years after beginning LAMP, only Boston Light Station, the oldest light in the United States, remained under the watch of a U.S. Coast Guard crew. At the urging of Sen. Edward Kennedy of Massachusetts, Congress passed a resolution in 1989 that Boston Light Station would remain permanently manned.[10]

Gradually, a few people began to realize that, as lighthouses began to shut down and people left after automation, the final act in a part of U.S. maritime heritage was passing. West Coast residents, for example, noticed the abandonment of the difficult light at St. George reef and the removal of the tower at Mile Rock light at the

The Pensacola Light Station in 1892—the first light built by the federal government on the Florida Gulf Coast and first lit in 1825. Still an active aid to navigation, the former light station is now located on a naval air station.

entrance to the Golden Gate. A helicopter platform constructed to service an automated beacon now stood in place of the tower. Some structures were left to vandals, others destroyed.

The love affair of Americans for their lighthouses began to assert itself as many from all walks of life rallied to save the structures. Such organizations as the U.S. Lighthouse Society, the Great Lakes Lighthouse Keepers Association, and others began work to save the lights. Lighthouses are also preserved in some state and national parks. People readily stay in semi-isolated locations to be "lighthouse keepers" and voluntarily donate their time and work to preserve lighthouses so that others might enjoy them. Today there are a number of lighthouses maintained as museums. For those wishing to view the actual equipment used by keepers of the U.S. Lighthouse Service and U.S. Coast Guard, the museum with the largest display of material is at the Shore Village Museum, Rockland, Maine, and is headed by Ken Black, a retired U.S. Coast Guard Chief Warrant Officer with aids to navigation experience.

Eventually anyone who ever served in the old U.S. Lighthouse Service will be gone, as will those who served aboard light stations as members of the U.S. Coast Guard. All the organizations that concern themselves with the preservation of lighthouses and related equipment operate on extremely tight budgets. This author encourages anyone who feels the history of aids to navigation in this country must be preserved to contact those working in the field. If it were not for the work of the various organizations and museums, perhaps another important part of the maritime heritage of the United States would be lost.

MAPS

✳

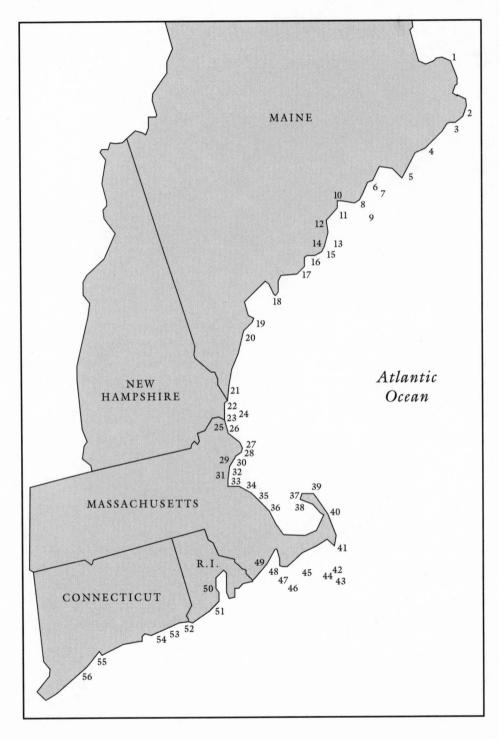

Principal Lighthouses of the Northeast Atlantic Coast

1. St. Croix River
2. West Quoddy Head
3. Libby Island
4. Moose Peak
5. Petit Manan
6. Bear Island
7. Baker's Island
8. Bass Harbor Head
9. Mount Desert Rock
10. Dice's Head
11. Saddleback Ledge
12. Owl's Head
13. Matinicus Rock
14. Whitehead
15. Monhegan Island
16. Franklin Island
17. Pemaquid Point
18. Sequin Island
19. Portland Head
20. Cape Elizabeth
21. Boon Island
22. Portsmouth Harbor
23. Whale's Back
24. Isles of Shoals
25. Newburysport Harbor
26. Ipswich
27. Annisquam Harbor
28. Cape Ann
29. Baker's Island
30. Derby Wharf
31. Marblehead
32. Long Island Head
33. Boston
34. Minots Ledge
35. Scituate
36. Plymouth
37. Race Point
38. Long Point
39. Cape Cod
40. Nauset Beach
41. Chatham Harbor
42. Nantucket
43. Sankaty Head
44. Brant Point
45. Cape Poge
46. Gay Head
47. Tarpaulin Cove
48. Butler Flats
49. Clark's Point
50. Beavertail
51. Point Judith
52. Windmill Point
53. Morgan's Point
54. New London
55. New Haven Harbor
56. Black Rock Harbor

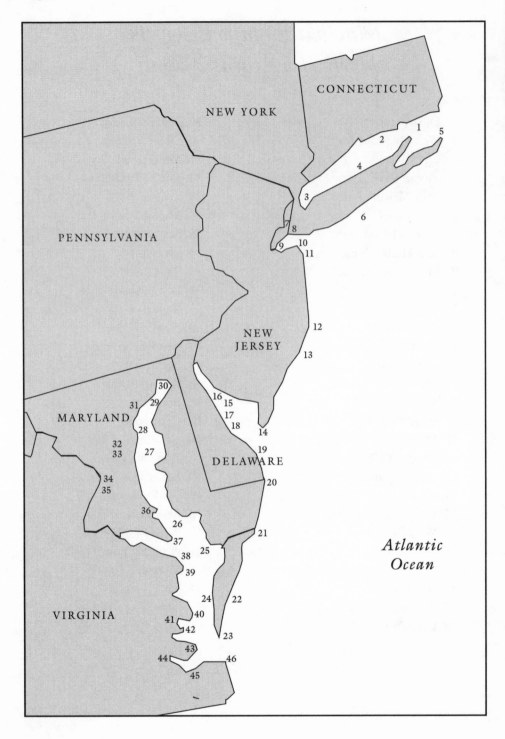

Principal Lighthouses of the Middle Atlantic Coast

1. Little Gull Island
2. Falkner's Island
3. Sands Point
4. Eaton's Neck
5. Montauk Point
6. Fire Island
7. Robbins Reef
8. Statue of Liberty
9. Fort Tompkins
10. Sandy Hook
11. Navesink
12. Barnegat
13. Absecon
14. Cape May
15. Cross Ledge
16. Ship John Shoal
17. Fourteen Foot Bank
18. Brandywine Shoal
19. Cape Henlopen
20. Fenwick Island
21. Assateague Island
22. Hog Island
23. Cape Charles

24. Pungoteague
25. Jane's Island
26. Hooper's Strait
27. Sharp's Island
28. Thomas Point
29. Pool's Island
30. Turkey Point
31. Fort Carroll
32. Jones Point
33. Ft. Washington
34. Upper Cedar Point
35. Lower Cedar Point
36. Drum Point
37. Point Lookout
38. Smith's Point
39. Windmill Point
40. Wolftrap
41. New Point Comfort
42. Thimble Shoals
43. Old Comfort Point
44. Point of Shoals
45. Craney Island
46. Cape Henry

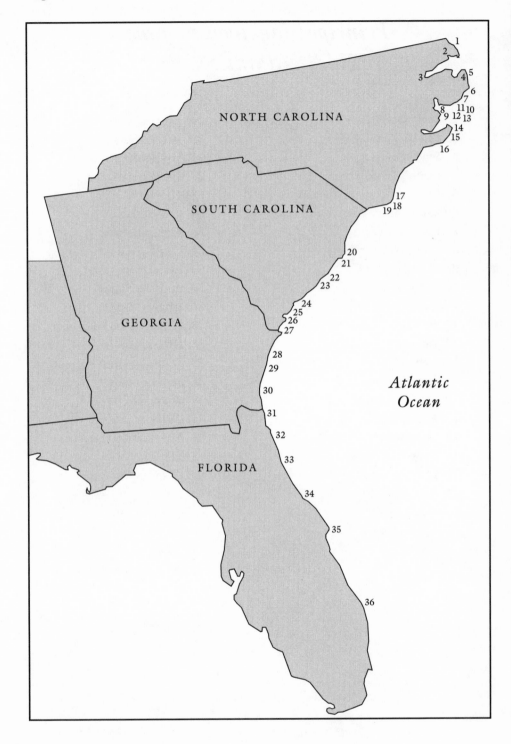

Principal Lighthouses of the South Atlantic Coast

1. Currituck Beach
2. Wade's Point
3. Roanoke River
4. Roanoke Marshes
5. Bodie Island
6. Horseshoe Shoal
7. Long Point Shoal
8. Pamplico Point Shoal
9. Brant Island Shoal
10. Ocracoke
11. Cape Hatteras
12. NW Point Royal Shoal
13. SW Point Royal Shoal
14. Beacon Island
15. Harbor Island Bar
16. Cape Lookout
17. Bald Head
18. Cape Fear
19. Federal Point
20. Georgetown
21. Cape Romain
22. Fort Sumter
23. Charleston
24. Combahee Bank
25. Hunting Island
26. Castle Pinckeney
27. Tybee
28. Sapelo Island
29. St Simon's
30. Little Cumberland Island
31. Amelia Island
32. St. John's River
33. St. Augustine
34. Ponce de Leon Inlet
35. Cape Canaveral
36. Juniper Inlet

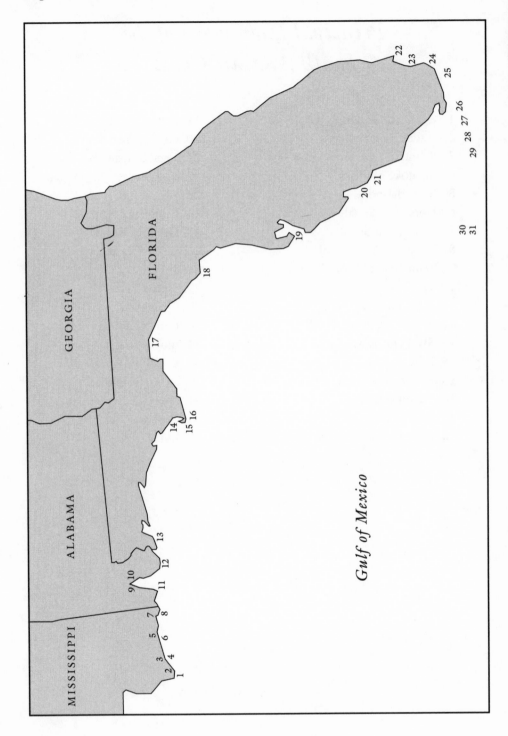

Principal Lighthouses of the East Gulf Coast

1. St. Joseph Island
2. Pass Christian
3. Merrill's Shell Bank
4. Cat Island
5. Biloxi
6. Ship Island
7. East Pascagoula River
8. Round Island
9. Choctaw Point
1O. Battery Gladden
11. Sand Island
12. Mobile Point
13. Pensacola
14. St. Joseph Bay
15. Cape San Blas
16. Cape St. George

17. St. Marks
18. Cedar Keys
19. Egmont Key
20. Gasparilla Island
21. Sanibel Island
22. Cape Florida
23. Fowey Rocks
24. Carysfort Reef
25. Alligator Reef
26. Sombrero Key
27. American Shoal
28. Key West
29. Sand Key
30. Loggerhead Key
31. Dry Tortugas

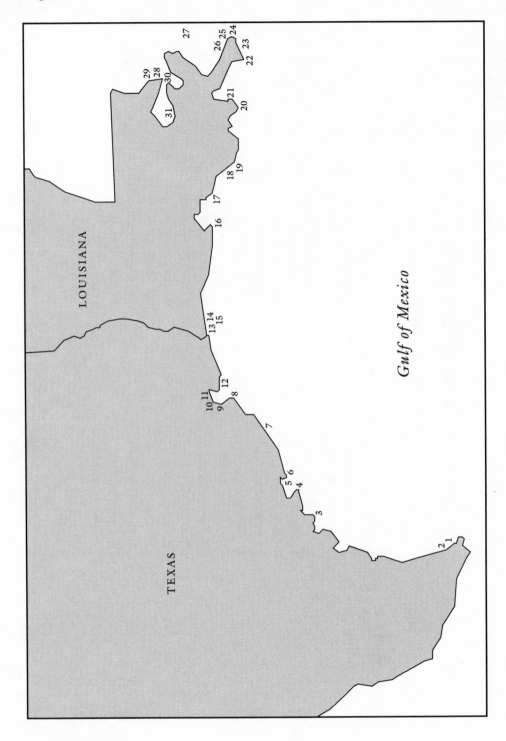

Principal Lighthouses of the West Gulf Coast

1. Point Isabel
2. Brazos Santiago
3. Aransas Pass
4. Matagorda
5. Swash
6. Decros Point
7. Brazos River
8. Fort Point
9. Halfmoon Shoals
10. Red Fish Bar
11. Cloppers Bar
12. Galveston Bay
13. Sabine Pass
14. Sabine Pass East Jetty
15. Sabine Bank
16. Trinity Shoals

17. Southwest Reef
18. Point au Fer
19. Point au Fer Reef
20. Timbalier Island
21. Barataria Bay
22. Southwest Pass
23. South Pass
24. Frank's Island
25. Pass a l'Outre
26. Balise Island
27. Chandeleur Island
28. Lake Borgne
29. East Rigolets
30. The Rigolets
31. Bayou St. John

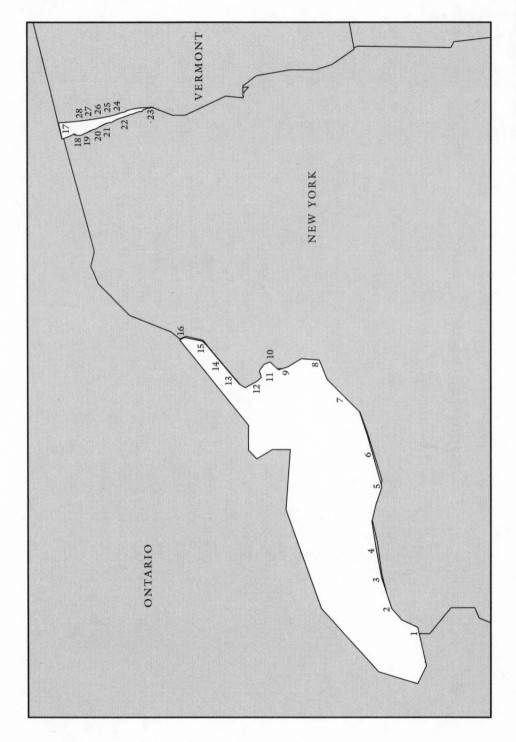

Principal Lighthouses of
Lake Ontario and Lake Champlain

1. Fort Niagara
2. Olcott
3. Thirty Mile Point
4. Oak Orchard
5. Rochester Harbor
6. Sodus Bay
7. Oswego
8. Salmon River
9. Stony Point
10. Sacketts Harbor
11. Galloo Island
12. Tibbetts Point
13. Rock Island
14. Sunken Rock

15. Oswegatchee
16. Crossover Island
17. Isle La Motte
18. Cumberland Head
19. Plattsburg Breakwater
20. Southwest Breakwater
21. Northeast Breakwater
22. Split Rock
23. Crown Point
24. Juniper Island
25. South Breakwater
26. Middle Breakwater
27. North Breakwater
28. Colchester Reef

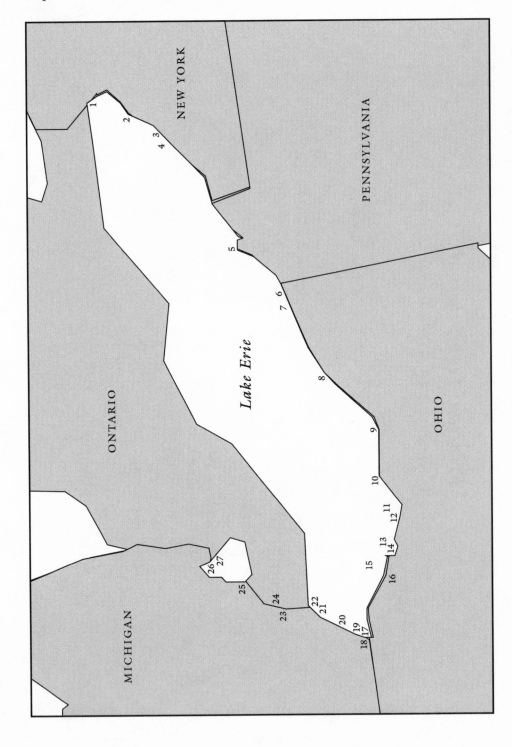

Principal Lighthouses of
Lake Erie

1. Buffalo
2. Otter Creek
3. Silver Creek
4. Barcelona
5. Presque Isle
6. Conneaut River
7. Ashtabula
8. Fairport
9. Cleveland
10. Black River
11. Cunningham Harbor
12. Huron River
13. Cedar Point
14. Sandusky

15. South Bass Island
16. Port Clinton
17. Maumee Bay
18. Turtle Island
19. Toledo Harbor
20. Monroe
21. Gibralter
22. Detroit River
23. Mamajudd Island
24. Grassy Island
25. Windmill Point
26. St. Clair Flats Canal Range
27. Lake St. Clair

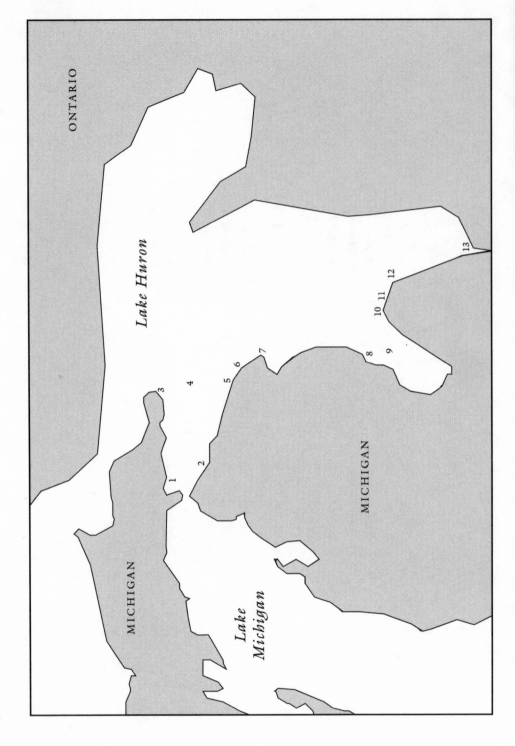

Principal Lighthouses of
Lake Huron

1. Bois Blanc Island
2. Cheboygan
3. Detour
4. Spectacle Reef
5. Forty Mile Point
6. Presque Isle Harbor
7. Thunder Bay Island

8. Tawas
9. Charity Island
10. Port Austin
11. Point aux Barques
12. Harbor Beach
13. Fort Gratiot

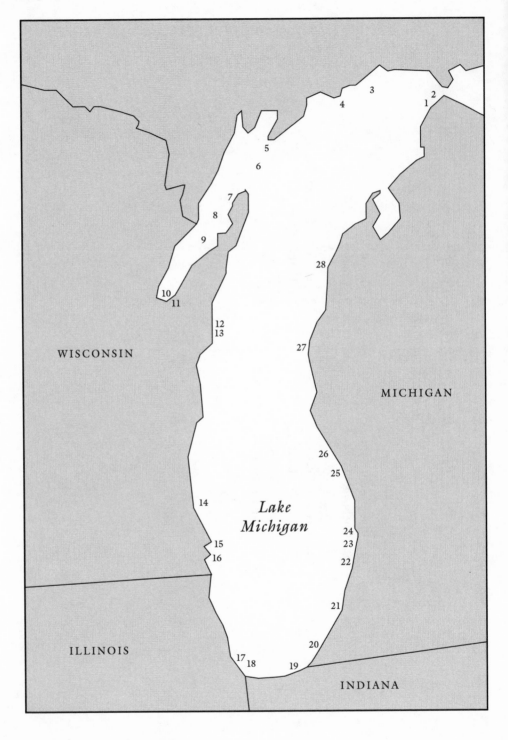

Principal Lighthouses of
Lake Michigan

1. Ile aux Galets
2. Gray's Reef
3. Lansing Shoal
4. Seul Choix Point
5. Poverty Point
6. Pottawatomie
7. Eagle Bluff
8. Green Island
9. Sherwood Point
10. Green Bay Harbor
11. Tail Point
12. Rawleys Point
13. Manitowoc
14. Milwaukee
15. Wind Point
16. Racine Reef
17. Chicago Harbor of Refuge
18. Chicago Harbor
19. Michigan City East Pierhead
20. New Buffalo
21. St. Joseph
22. South Haven Pierhead
23. Kalamazoo
24. Holland Harbor South Pierhead
25. Grand Haven South
    Pierhead Inner
26. Muskegon South Pierhead
27. Big Sable
28. Point Betsie

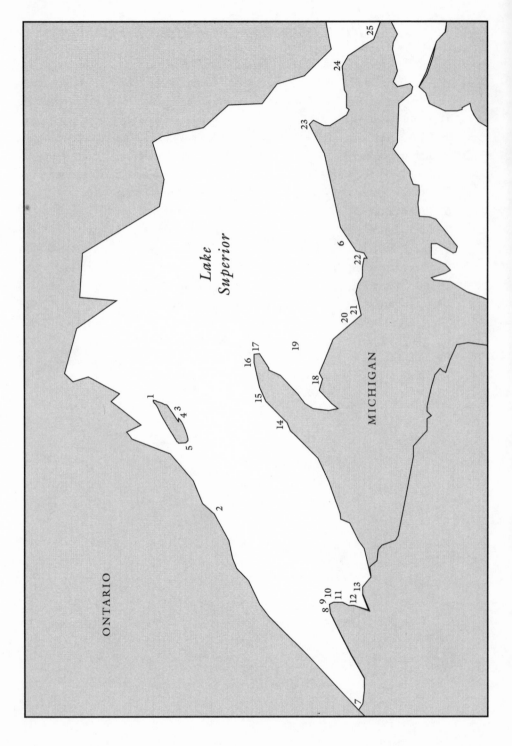

Principal Lighthouses of
Lake Superior

1. Passage Island
2. Split Rock
3. Rock Harbor
4. Menagerie Island
5. Rock of Ages
6. Au Sable
7. Duluth Harbor
8. Sand Island
9. Devil's Island
10. Outer Apostle Island
11. Michigan Island
12. Chequamegon Point
13. La Pointe

14. Calumet
15. Eagle River
16. Copper Harbor
17. Manitou Harbor
18. Huron Island
19. Stannard Rock
20. Granite Island
21. Marquette
22. Grand Island
23. Whitefish Point
24. Point Iroquis
25. Round Island

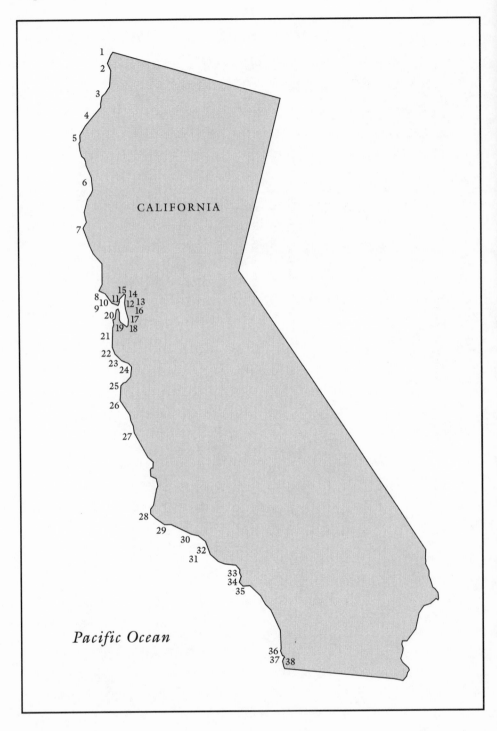

CALIFORNIA

Pacific Ocean

Principal Lighthouses of California

1. St. George's Reef
2. Crescent City
3. Trinidad Head
4. Humboldt Harbor
5. Cape Mendocina
6. Point Cabrillo
7. Pt. Arena
8. Point Reyes
9. S.E. Farallon Island
10. Point Bonita
11. Lime Point
12. Angel Island
13. Southhampton Shoal
14. East Brother Island
15. Carquinex Strait
16. Oakland Harbor
17. Yerba Buena Island
18. Alcatraz
19. Fort Point
20. Mile Rock
21. Point Montara
22. Pidgeon Point
23. Ano Nuevo
24. Santa Cruz
25. Point Pinos
26. Point Sur
27. Piedras Blancas
28. Point Arguello
29. Point Conception
30. Santa Barbara
31. Anacapa Island
32. Port Hueneme
33. Los Angeles Harbor
34. Point Vincent
35. Point Fermin
36. New Point Loma
37. Old Point Loma
38. Ballast Point

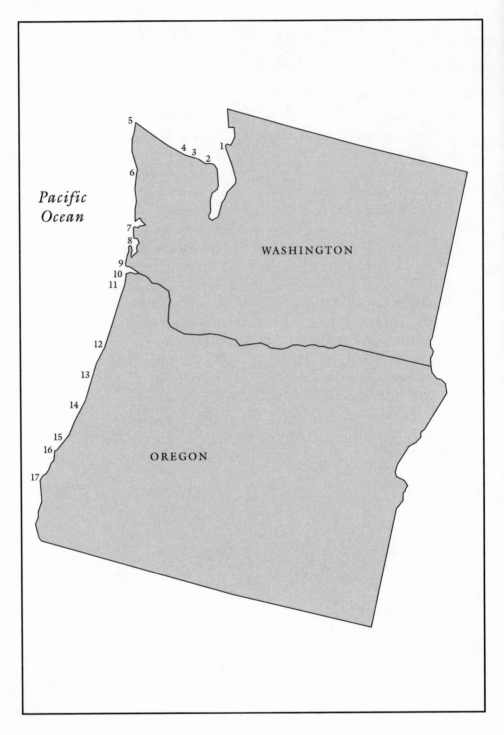

Pacific
Ocean

WASHINGTON

OREGON

Principal Lighthouses of
Pacific Northwest

1. Smith Island
2. Whidbey
3. Dungeness
4. Ediz Hook
5. Cape Flattery
6. Destruction Island
7. Gray's Harbor
8. Willapa Bay
9. North Head
10. Cape Disappointment
11. Tillamook Rock
12. Cape Meares
13. Yaquina Head
14. Heceta Head
15. Umpqua River
16. Cape Arago
17. Cape Blanco

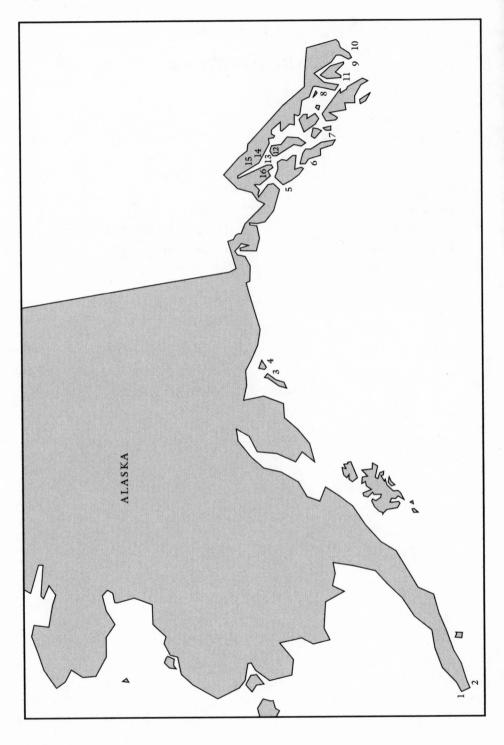

ALASKA

Principal Lighthouses of Alaska

1. Cape Sarichef
2. Scotch Cap
3. Cape Hinchinbrook
4. Cape St. Elias
5. Cape Spencer
6. Sitka Bay Beacon
7. Cape Decision
8. Lincoln Rock

9. Tree Point
10. Mary Island
11. Guard Island
12. Southeast Five Fingers Island
13. Point Sherman
14. Point Retreat
15. Eldred Rock
16. Sentinel Island

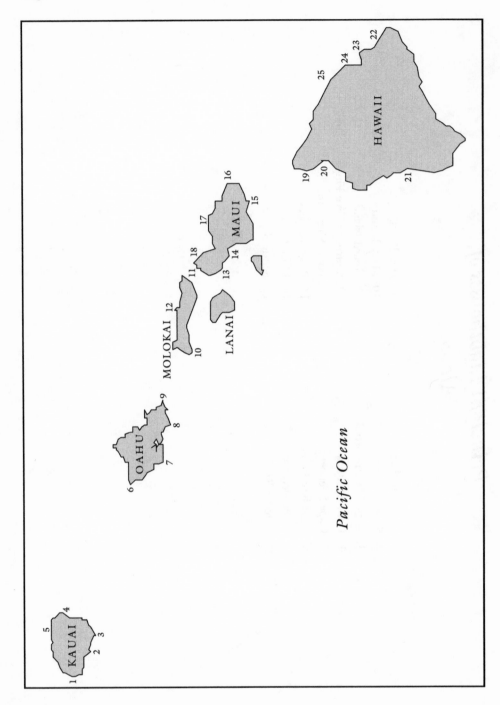

Principal Lighthouses of
Hawaii

1. Kakole
2. Hanapepe
3. Makahuena Point
4. Kahala Point
5. Kilauea
6. Kaena Point
7. Barbers Point
8. Diamond Head
9. Makapuu Point
10. Kalaeoka Lauu
11. Hawea Point
12. Molokai
13. Lahaina

14. Maalaea
15. Kanahene Point
16. Hanamanioa
17. Pauwela Point
18. Nakalele Point
19. Mahukina
20. Kawaihae
21. Napoopoo
22. Waiakea
23. Paukaa Point
24. Pepeekeo
25. Laupahoehoe

NOTES

1. A DIM BEACON

1. George R. Putnam, *Lighthouses and Lightships of the United States* (Boston: Houghton Mifflin Company, 1917), 288.
2. Ibid., 290.
3. Dudley Whitney, *The Lighthouse* (New York: New York Graphic Society, 1975), 32, 36.
4. Ibid., 36.
5. Francis Ross Holland, Jr., *America's Lighthouses: An Illustrated History* (Mineola, N.Y.: Dover Publications, 1988, reprint). Originally published: Rev. ed. Brattleboro, Vt.: Stephen Greene Press, 1981, 1972), 4 (all page numbers are keyed to the reprint edition).
6. Putnam, *Lighthouses and Lightships,* 259–60.
7. Holland, *America's Lighthouses,* 5–6.
8. Trinity House was a corporation chartered by Henry VIII for furthering British commerce and navigation and establishing lighthouses. Ibid., 6; Whitney, *The Lighthouse,* 17.
9. Sarah C. Gleason, *Kindly Lights: A History of Lighthouses of Southern New England* (Boston: Beacon Press, 1991), 4–6.
10. George Weiss, *The Lighthouse Service: Its History, Activities and Organization* (Baltimore: Johns Hopkins University Press, 1926), 1–2.
11. There have been suggestions of earlier lighthouses located in what is now the United States, namely in the Florida area. At present, however, most writers cite the Boston Light Station as the earliest. For some discussion on the possibility of other earlier locations, see *Journal of Andrew Ellicott, Late Commissioner on Behalf of the United States* (Philadelphia: Budd Hartam, 1803); Holland, *America's Lighthouses,* 8, 120–21.
12. Ibid., 9.
13. U.S. Coast Guard, *Historically Famous Lighthouses* (Washington, D.C.: U.S. Government Printing Office, 1957), 33.
14. Weiss, *Lighthouse Service,* 2–3.
15. Ibid., 4; Arnold Burges Johnson, *The Modern Light-house Service* (Washington, D.C.: U.S. Government Printing Office, 1890), 13–14.
16. Holland, *America's Lighthouses,* 27; Scheina, "Lighthouse Towers," 4.
17. Holland, *America's Lighthouses,* 27.

18. Ibid.; Johnson, *Modern Light-house,* 14–15.
19. Ibid.
20. Holland, *America's Lighthouses,* 27.
21. Ibid., 34.
22. Johnson, *Modern Light-house,* 14–15; Weiss, *Lighthouse Service,* 5.
23. Scheina, "Lighthouse Towers," 4–5.
24. Johnson, *Modern Light-house,* 15–17.
25. Ibid., 17.
26. Holland, *America's Lighthouses,* 30.
27. Quoted in Johnson, *Modern Light-house,* 18.
28. Holland, *America's Lighthouses,* 30.
29. Truman R. Strobridge, *Chronology of Aids to Navigation and the Old Lighthouse Service, 1716–1939* (Washington, D.C.: U.S. Coast Guard, 1974), 11.
30. Ibid.; Johnson, *Modern Light-house,* 16–20.
31. Johnson, *Modern Light-house,* 20.
32. Ibid., 21.
33. Ibid., 20–23; Holland, *America's Lighthouses,* 35.
34. Johnson, *Modern Light-house,* 20.
35. Holland, *America's Lighthouses,* 30–31, 32.
36. William James Morgan, et al., *Autobiography of Rear Admiral Charles Wilkes, U.S. Navy, 1798–1877* (Washington, D.C.: U.S. Department of the Navy, 1978), 317.
37. Ibid., 28.
38. Bill Caldwell, *Lighthouses of Maine* (Portland, Me.: Gannett Books, 1986), 202.
39. Dennis L. Noble and Ralph E. Eshelman, "A Lighthouse for Drum Point," *Keeper's Log* (Spring 1987): 4.
40. Whitney, *The Lighthouse,* 20–21.
41. Ibid., 21.
42. Ibid., 21–22.
43. Robert L. Scheina, "The Evolution of the Lighthouse Tower," in *Lighthouses: Then and Now* (Washington, D.C.: U.S. Coast Guard, n.d.), 3–4.
44. USCG, *Historically Famous Lighthouses,* 35–37.
45. Ibid., 63.
46. Ibid.
47. Ibid., 64; Holland, *America's Lighthouses,* 11.
48. Ibid., 13; Sara C. Gleason, *Kindly Lights,* 1–2.
49. Richard W. Updike, "Winslow Lewis and the Lighthouses," *American Neptune* 28, no. 1 (January 1968): 31.
50. Ibid., 31–32.
51. Ibid., 31, 33.
52. Ibid., 34–35; Gleason, *Kindly Lights,* 39–42.
53. Ibid., 35. Holland notes that Lewis eventually replaced the green lens, but it took "twenty-five years before all were [removed]." Holland, *America's Lighthouses,* 15.
54. Updike, "Winslow Lewis," 35–36; Gleason, *Kindly Lights,* 47–48.
55. H.A.S. Dearborn to Samuel Smith, 12 May 1816. Letters Received, Record Group 26 (RG26), Records of the U.S. Coast Guard (RUSCG), National Archives Building (NAB), Washington, D.C.
56. Updike, "Winslow Lewis," 38–40.
57. Ibid., 40; Gleason, *Kindly Lights,* 50–51.

58. David Melville, *An Expose of Facts, Respectfully Submitted to the Government and Citizens of the United States, Relating to the Conduct of Winslow Lewis, of Boston, Addressed to the Hon. the Secretary of the Treasury* (Providence, R.I.: Miller and Hutchens, 1819).

59. Updike, "Winslow Lewis," 41–43.

60. Ibid., 43.

61. Ibid., 43–44. Lewis claimed to have built eighty lighthouses, but Updike could find proof for only forty structures. Updike, "Winslow Lewis," 44n.

62. Holland, *America's Lighthouses,* 16.

63. Updike, "Winslow Lewis," 46.

64. Ibid., 46–47; Robert Fraser, "I.W.P. Lewis: Father of America's Lighthouse System," *Keeper's Log* (Winter 1989): 10. Fraser believes that it was I.W.P. Lewis's efforts to point out his uncle's poor work that led to the investigation of 1851 and the eventual overhaul of the lighthouse system.

65. *The Journal* (Boston), 20 May 1850, 2.

66. Johnson, *Modern Light-house,* 49.

67. Remey and Putnam quotes appear in Updike, "Winslow Lewis," 48.

68. Georges A. Boutry, *Augustin Fresnel: His Time, Life, and Work, 1788–1827* (London, Eng.: John Murray, 1949), 5–7; Lindy Lindquist, "L'Homme de Genie: The Unknown Tale of How 'Fresnel's Fancies' Lit the Lighthouses of the World," *Oceans* (November 1974): 58.

69. Boutry, *Augustin Fresnel,* 8.

70. Lindquist, "L'Homme de Genie," 59; Wayne Wheeler, "Augustin Fresnel and His Magic Lantern," *Keeper's Log* (Winter 1987): 9–10.

71. Lindquist, "L'Homme de Genie," 60–61.

72. Boutry, *Augustin Fresnel,* 10–11.

73. Ibid., 20.

74. Quoted in ibid., 18; quoted in Lindquist, "L'Homme de Genie," 63.

75. Holland, *America's Lighthouses,* 21, 23.

76. Ibid., 23.

77. Dennis L. Noble and T. Michael O'Brien, *Sentinels of the Rocks: From "Graveyard Coast" to National Lakeshore* (Marquette: Northern Michigan University Press, 1979), 8.

78. Holland, *America's Lighthouses,* 33.

2. A Brightly Burning Light

1. Holland, *America's Lighthouses,* 36.

2. Ibid.

3. Johnson, *Modern Light-house,* 23.

4. Holland, *America's Lighthouses,* 36.

5. "Instructions for Light-Keepers of the United States: Stations with Two or More Keepers," in files of the Historian of the U.S. Coast Guard, U.S. Coast Guard Headquarters, Washington, D.C. (hereafter cited as "Historian"). In addition to excellent photographs, some of the individual light station files in the Historian's office duplicate material copied from the National Archives, such as clipping files and the superb "Description of Light Station." Yet other files contain general material on the U.S. Lighthouse Service.

6. Letters to Ninth District Inspector, 10 June 1863–27 May 1867, RUSCG, RG26, NAB; F. R. Holland, Jr., *The Aransas Pass Light Station: A History* (Corpus Christi, Tex: privately printed, 1976), 4.

7. David L. Cipra, *Lighthouses and Lightships of the Northern Gulf of Mexico* (Washington, D.C.: U.S. Department of Transportation, 1976), 34–35, 37; Holland, *America's Lighthouses,* 108.

8. *Reports of the Department of Commerce and Labor, 1909* (Washington, D.C.: U.S. Government Printing Office, 1910), 49.

9. In 1850 there were 1,536 aids to navigation of all types; by 1910 the number had risen to 11,661. Weiss, *The Lighthouse Service,* 19.

10. Ibid., 20.

11. Holland, *America's Lighthouses,* 37.

12. Ibid., 23; USCG, *Historically Famous Lighthouses,* 54.

13. For details of Putnam's life, see, George R. Putnam, *Sentinel of the Coasts: The Log of a Lighthouse Engineer* (New York: W. W. Norton, 1937), 1–120; Holland, *America's Lighthouses,* 37–38.

14. Noble and O'Brien, *Sentinels of the Rocks,* 8.

15. Mary Louise Clifford and J. Candace Clifford, *Women Who Kept the Lights: An Illustrated History of Female Lighthouse Keepers* (Williamsburg, Va.: Cypress Communications, 1993), 65–66; Patricia Harris, "Michigan City, Indiana's Only Lighthouse," *Keeper's Log* (Spring 1987): 24.

16. Putnam, *Lighthouses and Lightships,* 186.

17. Ibid., 188.

18. Ibid., 188. The Statue of Liberty was maintained as a lighthouse by the Lighthouse Board from 1886 to 1902. Putnam, *Lighthouses and Lightships,* 65.

19. Holland, *America's Lighthouses,* 25.

20. Quoted in Stephen H. Evans, *The United States Coast Guard, 1790–1915: A Definitive History* (Annapolis, Md.: Naval Institute Press, 1949), 187–88.

21. The route to the modern-day U.S. Coast Guard is complicated. The U.S. Revenue Cutter Service merged with the U.S. Life-Saving Service to form the U.S. Coast Guard on 20 January 1915. On 1 July 1939, the U.S. Coast Guard absorbed the U.S. Lighthouse Service, officially known as the Bureau of Lighthouses, and in 1942, the U.S. Coast Guard absorbed the Bureau of Marine Inspection and Navigation. Dennis L. Noble, *That Others Might Live: The U.S. Life-Saving Service, 1878–1915* (Annapolis, Md.: Naval Institute Press, 1994), 150–55; Robert Erwin Johnson, *Guardians of the Sea: History of the United States Coast Guard, 1915 to the Present* (Annapolis, Md.: Naval Institute Press, 1987), 162.

22. For some of the machinations of the establishment of the U.S. Coast Guard, see, Noble, *That Others Might Live,* 149–53.

23. Holland, *America's Lighthouses,* 38. Putnam died on 2 July 1953 and is buried at Dorset, Vermont.

24. Compare this to the U.S. Life-Saving Service, whose employees never received a retirement benefit. See Noble, *That Others Might Live,* especially 149.

25. Putnam, *Lighthouses and Lightships.* The other work is Holland, *America's Lighthouses.*

26. Holland, *America's Lighthouses,* 38; Robert H. Macy, "The Consolidation of the Lighthouse Service with the Coast Guard," U.S. Naval Institute *Proceedings* 66, no. 1 (January 1940): 58.

27. Johnson, *Guardians of the Sea,* 162.

28. Ibid.

29. Ibid.; Macy, "Consolidation of the Lighthouse Service," 59.

30. Ibid., 163–64; ibid., 58.

31. Johnson, *Guardians of the Sea,* 164.

32. Ibid., 163–64.

3. Tales of Seven Beacons

1. Unless otherwise noted, all material on the various types of light stations comes from Historian; Candace Clifford, *Inventory of Historic Light Stations* (Washington, D.C.: National Park Service, 1994), passim; Johnson, *Modern Light-House Service,* passim; Holland, *America's Lighthouses,* passim; Putnam, *Lighthouses and Lightships,* passim; George R. Putnam, "Beacons of the Sea: Lighting the Coasts of the United States," *National Geographic* 24, no. 1 (January 1913): passim; Scheina, "Lighthouse Towers," passim.

2. Quoted in M. A. DeWolfe Howe, *The Humane Society of the Commonwealth of Massachusetts: An Historical Review* (Boston: Humane Society, 1918), 89. The Massachusetts Humane Society, formed in Boston in 1785, was modeled on a similar British organization.

3. William A. Baker, *A History of the Boston Marine Society, 1742–1967* (Boston: Boston Marine Society, 1968), 73.

4. Baker, *History of Boston Marine Society,* 73.

5. Site file for Cape Cod, RG26, NAB.

6. Benjamin Lincoln (1733–1810) apparently received his post "as a reward for his services in the War for Independence, and for crushing Shay's Rebellion in 1787. . . . [During his tenure,] Lincoln was directly responsible for far more lighthouses than any other superintendent." Peggy A. Albee, et al., *Highland Lighthouse and Keeper's Dwelling: Historic Structure Report: North Truro, Massachusetts* (Lowell, Mass.: National Park Service, North Atlantic Region, 1994), 8.

7. Tench Coxe to Gen. Benjamin Lincoln, 11 June 1796. Correspondence, Series 17, RG26, NAB.

8. Ibid., 20.

9. Ibid., 25–26.

10. Ibid., 27.

11. Quoted in ibid., 29.

12. Albee, *Highland Lighthouse,* 30.

13. Dearborn to Albert Gallatin, 27 June 1811. Correspondence, RG26, NAB.

14. Lewis to Dearborn, June 1812. Correspondence, RG26, NAB; Lighthouse Deeds and Contracts, RG26, NAB.

15. Annual Report, 1828, Box 2, Lighthouse Annual Reports, 1820–1853, Series 6 (Annual Reports), RG26, NAB.

16. Cape Cod clipping file, RG26, NAB.

17. Clipping file.

18. Ibid.

19. *Journal of the Light-House Board,* 20 May 1851–1 January 1908, Series 1, RG26, NAB.

20. Ibid.

21. Albee, *Highland Light,* 50–53. The actual dates of construction of the principal keeper's dwelling and the costs for the entire project are difficult to determine because records were destroyed.

22. Ibid., 54.

23. Clipping file.

24. Albee, *Highland Light,* 105.

25. Ibid., 118–25.

26. Ibid., 130.

27. News releases, National Park Service, Cape Cod National Seashore, 1 June 1994 and 8 March 1996, located in the lighthouse files of the U.S. Lighthouse Society (USLHS), San

Francisco, California. As the final draft of this book was completed (July 1996), work was progressing on the move.

28. *American State Papers, Class 4,* vol. 1, *Commerce and Navigation, 1st–13th Congress, 1789–1815* (Washington D.C.: U.S. Government Printing Office, 1832), 265–67, 299.

29. Lighthouse Deeds and Contracts, RG26, NAB; Lighthouse Letters, RG26, NAB.

30. Lighthouse Letters from Lighthouse Superintendents; Letters Received, RG26, NAB.

31. Lighthouse Letters, RG26, NAB.

32. Ibid.; Cape Hatteras clipping file, RG26, NAB.

33. Lighthouse Letters; Lighthouse Deeds and Contracts; Letters Received, RG26, NAB.

34. Samuel Treadwell to William Miller, 2 and 25 November 1798, Lighthouse Letters from Lighthouse Superintendents; Lighthouse Deeds and Contracts, RG26, NAB.

35. F. Ross Holland, Jr., *A History of the Cape Hatteras Light Station: Cape Hatteras National Seashore, North Carolina* (Washington, U.S.: National Park Service, Division of History, 30 September 1968), 9. Holland's work is a model that should be followed by those wishing to document the lighthouses of the United States. This study covers the Cape Hatteras Light Station, Diamond Shoals Lightship, the Cape Hatteras Beacon, the Hatteras Inlet Light Station, and the Shell Castle Beacon. Holland's detailed work should be consulted by anyone wishing to understand the history of the aids to navigation in the Cape Hatteras area.

36. Quoted in ibid., 10.

37. Quoted in ibid., 11.

38. Ibid., 12, 14.

39. Ibid., 17–18.

40. Ibid., 24–25.

41. Treadwell to Samuel H. Smith, Lighthouse Letters from Lighthouse Superintendents; Lighthouse Letters, RG26, NAB.

42. Lighthouse Letters; site file, Cape Hatteras Light Station, RG26, NAB.

43. Holland, *Cape Hatteras,* 38.

44. Lighthouse Letters, RG26, NAB.

45. R.H.J. Blount to Thomas Corwin, 7 January 1851, Lighthouse Letters, RG26, NAB; Holland, *Cape Hatteras,* 49–50.

46. Lighthouse Letters, RG26, NAB; Isherwood went on to become engineer in chief of the U.S. Navy and first chief of the service's Bureau of Steam Navigation. He is noted for his innovations in ship design and machinery. *Dictionary of American Biography,* vol. 9, 515–16.

47. U.S. Lighthouse Board, *Compilation of Public Documents and Extracts from Reports and Papers Relating to Light-houses, Light-vessels, and Illuminating Apparatus, and to Beacons, Buoys, and Fog Signals, 1789 to 1871* (Washington, D.C.: U.S. Government Printing Office, 1871), 882; *Report of the Light-House Board to the Secretary of the Treasury in Answer to a Resolution of the Senate of February 1, 1858, Calling Upon the Department for Information in Regard to the Expense of Erecting Light-Houses, &c* (Washington, D.C.: U.S. Government Printing Office, 1858), 44–45.

48. U.S. Lighthouse Board, *Compilation of Public Documents . . . 1789 to 1871,* 738, 743, 753–54; 32d Cong., 1st sess., H. Exec. Doc. 114 (ser. no. 648): 17, 23.

49. Clipping file; *Report of the Light-House Board . . . Regard to Expense of Erecting Light-Houses,* 98–99.

50. Holland, *Cape Hatteras,* 67–68.

51. Lighthouse Letters, RG26, NAB.

52. Quoted in Holland, *Cape Hatteras,* 76.

53. Ibid., 83–86; clipping file.

54. Holland, *America's Lighthouses,* 116; clipping file.

55. Clipping file.

56. Dawson Carr, *The Cape Hatteras Lighthouse: Sentinel of the Shoals* (Chapel Hill: The University of North Carolina Press, 1991), 81, 97.

57. Ibid., 93.

58. Ibid., 99–100.

59. Ibid., 105–6.

60. Ibid., 115–16.

61. Ibid., 123–30, 132.

62. Drum Point clipping file, RG26, NAB.

63. Unless otherwise noted, all material on screwpile lights comes from Layne Bergin, "Screwpile Lighthouses: From Britain to the Bay," *Keeper's Log* (Spring 1987): 10–15.

64. Unless otherwise noted, all material on the Drum Point Light Station comes from Noble and Eshelman, "A Lighthouse for Drum Point," 2–9.

65. Clifford and Clifford, *Women Who Kept the Lights,* 1; Holland, *America's Lighthouses,* 139.

66. All material on Meade's report is found in a photocopy of a letter to Col. J. J. Albert, Topographical Engineers, Washington, D.C., 27 August 1853, located in the Sand Key Light Station file, USLHS.

67. Sand Key file, Historian.

68. The description of Sand Key lighthouse is contained in "Descriptive Lists of Lighthouse Stations, 1858–1889, 1876–1939, 1876–1938," passim., RG26, NAB.

69. Sand Key file, Historian.

70. Ibid.; letter, Capt. B. W. Hadler, chief, Aids to Navigation and Waterways Management Branch, Seventh Coast Guard District, Miami, Fla., to the author, 16 May 1996.

71. Ibid.

72. Putnam, "Beacons of the Sea," 19.

73. Unless otherwise noted, all material on the Spectacle Reef lighthouse is found in the Spectacle Reef file, Historian. Much of the file contains material from the clipping file at the National Archives.

74. Undated, unnumbered paper in Spectacle Reef file, Historian.

75. Putnam, "Beacons of the Sea," 19.

76. Clifford, *Inventory of Historic Light Stations,* 202.

77. Holland, *America's Lighthouses,* 153.

78. Unless otherwise noted, all material on the construction of Tillamook Rock light comes from a photocopy of the clipping file in the Tillamook Rock file, Historian.

79. The author can vividly remember the sinking feeling in his stomach when, as a "boot" at his first duty station in the U.S. Coast Guard, he was told to jump a distance of only three to four feet from a boat to a ladder on a lighthouse pier and in seas of only one to two feet.

80. James A. Gibbs, Jr., *Tillamook Light* (Portland, Ore.: Binfords and Mort, 1953), 60–61. Gibbs served a short period of time during the 1940s on Tillamook Light Station while in the U.S. Coast Guard. He indicates that the station could be used as a punishment tour. He was the only military man on the rock, the other three men were civilian keepers. C. H. McClelland gives Ballantyne's name as "Charles C.," the official reports simply give his first initial of "A," while on page 6 of McClelland's, there is an extract of the journal kept by Ballantyne and he is again identified as "A. Ballantyne." C. H. McClelland, "Terrible Tilly," *Keeper's Log* (Summer 1987): 4, 6.

81. Putnam, "Beacons of the Sea," 23.

82. Work began on the St. George Reef Light Station in 1883, and the light was lit on 20 October 1892. It is on a rock only three hundred feet in diameter, located six miles out to sea. The foundation of the St. George Reef light tower is a pier in an irregular oval shape, eighty-six feet in diameter, faced with cut granite. The tower is also constructed of granite; the smallest block weighs seventeen tons. The light station stands 144 feet above sea level. This was the most expensive to build up to that time, costing $704,633.78. Holland, *America's Lighthouses,* 171–72; Dennis L. Noble, *Southwest Pacific: A Brief History of U.S. Coast Guard Operations* (Washington, D.C.: U.S. Coast Guard, 1989), 2–3; Gibbs, *Tillamook Light,* 64.
83. Quoted in Gibbs, *Tillamook Light,* 64.
84. Ibid., 75; Dennis L. Noble, *The Coast Guard in the Pacific Northwest* (Washington, D.C.: U.S. Coast Guard, 1988), 4, 7.
85. Photocopy of logbook entry in Tillamook Rock file, Historian.
86. All material on the various dispositions of the light comes from photocopied clippings in the Tillamook Rock file, Historian.
87. The current usage in Hawai'i of Makapuu Point is Makapu'u. I have left the spelling as it was when the light station was in operation.
88. Love Dean, *The Lighthouses of Hawai'i* (Honolulu: University of Hawaii Press, 1991), 38–39.
89. Ibid., 39–40.
90. Ibid., 40; U.S. House of Representatives, 59th Cong., 1st sess., report no. 159, 1.
91. Descriptive Lists of Lighthouse Stations, "Makapuu Point," 14 September 1910, 3, RG26, NAB.
92. Makapuu file, Historian.
93. "Description of Makapuu Point, 1910," 16–17.
94. Ibid., 4–6; Makapuu file, Historian.
95. Quoted in Dean, *Lighthouses of Hawai'i,* 42; Makapuu file, Historian.
96. Ibid., 42–43; ibid.
97. Ibid., 43; ibid.
98. Makapuu Point clipping file, RG26, NAB; "Description of Makapuu, 1910," 8–11, 18; "Description of Makapuu, 1916," 12.
99. Dean, *Lighthouses of Hawai'i,* 45; Department of Commerce, Lighthouse Service *Bulletin* 3, no. 18 (1 June 1925), 80.
100. Lighthouse Service *Bulletin* 3, no. 17 (1 May 1925), 76; Dean, *Lighthouses of Hawai'i,* 45.
101. Dean, *Lighthouses of Hawai'i,* 46.
102. Ibid., 50; Makapuu file, Historian.
103. Elinor DeWire, "The Bulging Eye of Makapu'u," *Keeper's Log* (Summer 1986): 4; Dean, *Lighthouses of Hawai'i,* 51–53.
104. Letter, with enclosures, MacKinnon Simpson, Hawaii Maritime Center, Honolulu, Hawaii, 25 May 1996, the author.

4. KEEPERS AND THEIR LONELY WORLD

1. Quoted in Noble and O'Brien, *Sentinels of the Rocks,* 9.
2. This chapter deals only with keepers of lighthouses; life on a lightship or buoy tender is discussed in chapters 6 and 7.
3. Putnam, *Lighthouses and Lightships,* 8–9. Worthylake, together with his wife and daughter, drowned on 3 November 1718.

4. 32d Cong., 1st sess. H. Exec. Doc. 55, 64.

5. All letters are located in "The Lighthouse Service, 1845," Collection MVF 178, Museum, U.S. Coast Guard Academy, New London, Connecticut.

6. Holland, *America's Lighthouses*, 44.

7. Ibid., 40.

8. Ibid., 44–45.

9. U.S. Lighthouse Board, *Instructions to Light-Keepers and Masters of Light-House Vessels* (Washington, D.C.: U.S. Government Printing Office, 1902), 27.

10. From an undated and untitled publication located in the Lighthouse Miscellaneous file, Historian.

11. Holland, *America's Lighthouses*, 46.

12. Wayne Wheeler, "The Keeper's New Clothes," *Keeper's Log* (Summer 1985): 11.

13. "Lighthouse Keepers and Assistants," various dates, RG26, NAB.

14. Holland, *America's Lighthouses*, 46; 32d Cong., 1st sess., H. Exec. Doc. 55, 64.

15. The author of this long poem is unknown. Pat Hall, *The Point Reyes Light* (Point Reyes, Calif.: Coastal Parks Association, 1979), 12.

16. U.S. Lighthouse Board, *Instructions to Light-Keepers* (Washington, D.C.: U.S. Government Printing Office, 1902), 27; U.S. Lighthouse Board, *General Orders* (Washington, D.C.: U.S. Government Printing Office, 1870), 8.

17. U.S. Lighthouse Establishment, *Instructions to Keepers: July 1881* (Washington, D.C.: U.S. Government Printing Office, 1881), 28.

18. Putnam, *Lighthouses and Lightships*, 248.

19. Ralph C. Shanks, Jr., and Janetta Thompson Shanks, *Lighthouses and Lifeboats on the Redwood Coast* (San Anselmo, Calif.: Costano Books, 1978), 86.

20. James A. Gibbs, Jr., *West Coast Lighthouses: A Pictorial History of the Guiding Lights of the Sea* (Seattle, Wash.: Superior Publishing Company, 1974), 126.

21. Caldwell, *Lighthouses of Maine*, 180–81.

22. Bernard J. Bretherton, "The Destruction of Birds by Lighthouses," *Osprey* (1902): 76–78, a clipping in Lighthouse Miscellaneous file, Historian.

23. Quoted in Noble and O'Brien, *Sentinels of the Rocks*, 9.

24. This story has been told in many books about lighthouses; probably the first time it appeared was in Putnam's, *Sentinels of the Lights*, 250. Putnam, unfortunately, did not document the material in his books. The quote is from one of the latest accounts of Alaskan lights; Shannon Lowery, *Northern Lights: Tales of Alaska's Lighthouses and Their Keepers* (Harrisburg, Pa.: Stackpole Books, 1992), 31–32.

25. Gibbs, *West Coast Lighthouses*, 112–13, 155. Gibbs noted that the keeper described was himself while he served in the U.S. Coast Guard.

26. Putnam, *Sentinels of the Lights*, 255.

27. Shanks and Shanks, *Lighthouses and Lifeboats on the Redwood Coast*, 240–41.

28. Lowery, *Northern Lights*, 33.

29. Gibbs, *West Coast Lighthouses*, 128, 143; Shanks and Shanks, *Lighthouses and Lifeboats on the Redwood Coast*, 53.

30. Dewey Livingston, "The Keepers of the Light," *Keeper's Log* (Winter 1991): 17, 18. For an excellent, detailed history of the Point Reyes Light Station, see Dewey Livingston and Dave Snow, *The History and Architecture of the Point Reyes Light Station* (Point Reyes, Calif.: National Park Service, 1990).

31. Shanks and Shanks, *Lighthouses and Lifeboats on the Redwood Coast*, 52.

32. Ibid., 56–59.

33. Gibbs, *Tillamook Rock*, 5.

34. Quoted in Caldwell, *Lighthouses of Maine,* 269–70.
35. Descriptive Lists of Lighthouses, Grand Island Lighthouse, 19 August 1909, RG26, NAB.
36. Journal of the Big Sable, Michigan, Light-Station, May 1881, RG26, NAB. Originally the station was named Big Sable, which was then changed to Au Sable.
37. Holland, *America's Lighthouses,* 48–49.
38. Noble and O'Brien, *Sentinels of the Rocks,* 11.
39. Quoted in Shanks and Shanks, *Lighthouses and Lifeboats on the Redwood Coast,* 80. The author can remember a shipmate informing him of a visit to a retired civilian lighthouse keeper's home. In his kitchen all the doors in the pantry were locked with padlocks, the keeper habitually withholding extra supplies, even for his own home.
40. Gleason, *Kindly Lights,* 129; Cipra, *Northern Gulf Coast,* 24.
41. Unless otherwise noted, the material on the October 1934 storm is from Sam Churchill, "The Day 'Terrible Tillie's' Light Nearly Died in a Sea of Terror," *Northwest Magazine* (3 December 1972): 6, 8.
42. *Lighthouse Service Bulletin* 4, no. 59 (1 November 1934), 191.
43. All material on Keeper Hanna's rescue is in the U.S. Life-Saving Service's *Annual Report of the Operations of the U.S. Life-Saving Service for the Fiscal Year Ending June 30, 1885* (hereafter cited as *USLSS, AR,* with appropriate year and page[s]) (Washington, D.C.: U.S. Government Printing Office, 1885), 42–44. Hanna's military Medal of Honor during the Civil War was received while he was a sergeant in Company B, 50th Massachusetts Infantry, on 4 July 1863, at Port Hudson, Louisiana; Hanna "voluntarily exposed himself to a heavy fire to get water for comrades in rifle pits." *The Medal of Honor of the United States Army* (Washington, D.C.: U.S. Government Printing Office, 1948), 143.
44. Shanks and Shanks, *Lighthouses and Lifeboats on the Redwood Coast,* 61.
45. Quoted in Noble, *Coast Guard in Pacific Northwest,* 7.
46. Ibid.
47. Letter, chief, Aids to Navigation Section, to commander, 13th CG District, 13 May 1952, in Tillamook file, Historian.
48. Ted Pedersen, as told to Ed Moreth, "Ted Pedersen: An Alaskan Lighthouse Keeper," *Keeper's Log* (Spring 1990): 19.
49. LuAnne Gaykowski Kozma, *Living at a Lighthouse: Oral Histories from the Great Lakes* (Detroit: Great Lakes Lighthouse Keepers Association/Harlo Printing, 1987), 23.
50. Putnam, *Sentinel of the Coast,* 243, 248.
51. *A History of Blacks in the Coast Guard from 1790* (Washington, D.C.: U.S. Coast Guard, n.d.), 14.
52. Truman R. Strobridge, "Blacks and Lights," *Shipmates* 3, no. 4 (Summer 1975): 15.
53. Quoted in ibid.
54. Quoted in ibid.
55. Quoted in ibid.
56. Ibid., 16.
57. Ibid., 17.
58. Quoted in ibid. For a short period during the Civil War, the U.S. Lighthouse Board permitted a lightship to be entirely manned by African Americans in South Carolina. Holland, *America's Lighthouses,* 41.
59. Strobridge, "Blacks and Lights," 17.
60. All material on the *Lammerlaw* rescue is found in *USLSS, AR, 1882,* 118–21.
61. All material on the wreck and rescue of the *San Benito* is found in Medals, U.S. Life-Saving Service, RG26, NAB.

62. Miller is not identified as being a Native American in the text of the annual report of the U.S. Life-Saving Service, but he is clearly so recognized in the written original reports of the rescue, ibid.

63. Clifford and Clifford, *Women Who Kept the Lights,* 169.

64. Ibid., 23.

65. The original towers were made of wood. Matinicus Rock file, Historian.

66. Abbie Burgess's letters are quoted in a number of works on lighthouses. This quote is from Clifford and Clifford, *Women Who Kept the Lights,* 25.

67. Ibid.

68. Ibid., 26–27.

69. Ibid., 28.

70. Caldwell, *Lighthouses of Maine,* 166.

71. Quoted in Clifford and Clifford, *Women Who Kept the Lights,* 28–29.

72. Snow apparently had a deep feeling for those who kept the lights; so much so that he used royalties realized from his lighthouse books to help finance a Christmas project. Snow would use his own airplane to drop Christmas gifts to many lighthouse keepers, thus earning the nickname "flying Santa Claus." Dorothy Holder Jones and Ruth Sexton Sargent, *Abbie Burgess, Lighthouse Heroine* (New York: Funk and Wagnalls, 1969), 190.

73. Sue Ellen Thompson, "The Light Is My Child," *Log of Mystic Seaport* 32, no. 3 (Fall 1980): 90.

74. Ida Lewis file, Historian.

75. Quoted in Clifford and Clifford, *Women Who Kept the Lights,* 97; Thompson, "Light Is My Child," 90–91; *USLSS, AR, 1881,* 87.

76. Clifford and Clifford, *Women Who Kept the Lights,* 94–95; Ida Lewis file.

77. Thompson, "The Light Is My Child," 92.

78. Ibid., 98.

79. In 1927, the U.S. Lighthouse Service removed the lens from the light and placed an automated light on a skeleton tower in front of the light, which was deactivated by the U.S. Coast Guard in 1963. The Newport Yacht Club purchased the lighthouse and maintains it as a private aid to navigation and a club. It is named the Ida Lewis Yacht Club. Clifford and Clifford, *Women Who Kept the Lights,* 98.

80. Ibid., 96–97; Ida Lewis file.

81. Quoted in Clifford and Clifford, *Women Who Kept the Lights,* 128.

82. Ibid., 127.

83. Quoted in ibid., 128.

84. New York *Times,* 5 March 1906, Section 3, 7.; Clifford Gallant Papers, U.S. Lighthouse Society, San Francisco, California.

85. All material on Laura Hecox is found in Clifford and Clifford, *Women Who Kept the Lights,* 121–25.

86. Gleason, *Kindly Lights,* 124.

5. Ghosts and the Places They Haunt

1. Elinor DeWire, *Guardians of the Lights: The Men and Women of the U.S. Lighthouse Service* (Sarasota, Fla.: Pineapple Press, 1995), 227.

2. Caldwell, *Lighthouses of Maine,* 57.

3. Ibid., 133.

4. Ibid., 154–55.

5. Ibid., 226.

6. Whitney, *The Lighthouse*, 159.

7. Elinor DeWire, "Specters on the Spiral Stairs," *Keeper's Log* (Winter 1986): 8.

8. Ibid., 10.

9. Whitney, *The Lighthouse*, 196; Execution Rock file, Historian.

10. Shanks and Shanks, *Lifeboats and Lighthouses*, 218.

11. Edward Rowe Snow, *Famous Lighthouses of America* (New York: Dodd, Mead, 1955), 115.

12. Ibid., 64.

13. Ibid.; DeWire, *Guardians of the Lights*, 252.

14. Edward Rowe Snow, *The Lighthouses of New England* (New York: Dodd, Mead, 1973), 222.

15. Whitney, *The Lighthouse*, 175.

16. Harlan Hamilton, *Lights and Legends: A Historical Guide to Lighthouses of Long Island Sound, Fishers Island Sound and Block Island Sound* (Stamford, Conn.: West Cove Publishing, 1987), 189–90.

17. Ibid., 77–78.

18. Whitney, *The Lighthouse*, 241.

19. Gibbs, *West Coast Lighthouses*, 49.

20. Shanks and Shanks, *Lighthouses and Lifeboats*, 218.

21. *The Alaskan* (Sitka), 27 April 1889, 1; 10 November 1888, 2.

22. DeWire, *Guardians of the Lights*, 233–34.

23. This light is not to be confused with the Yaquina Head lighthouse north of present-day Newport.

24. Ibid., 246–47.

25. Interview, executive petty officer, U.S. Coast Guard Aids to Navigation Team, Port Angeles, Washington, and Dennis L. Noble, 10 October 1995.

6. Lighthouses Go to Sea

1. Johnson, *Modern Lighthouse Service*, 42.

2. Willard Flint, *A History of U.S. Lightships* (Washington, D.C.: U.S. Coast Guard, 1993), 2; Robert F. Cairo, "Notes on Early Lightship Development," *U.S. Coast Guard Engineer's Digest*, 188 (July–August–September 1975): 4; Bernard C. Nalty and Truman R. Strobridge, "A Bright and Steadfast Light," U.S. Coast Guard Academy *Alumni Bulletin* 38, no. 6 (November–December 1975): 37–38.

3. Flint, *History of Lightships*, 2–3; Willard Flint, *Lightships of the U.S. Government: Reference Notes* (Washington, D.C.: U.S. Coast Guard, 1989), n.p. This massive work on lightships is the best single book for research on the lightships of the United States. Because much of the material within the publication is reproduced from the National Archives, using this book is much like visiting their offices. Anyone wishing to find information on lightships should consult this work. The only drawback to the book is that the pages are not numbered. Hereafter, the book will be cited as Flint, *Lightships of the U.S.*, without page numbers.

4. Holland, *America's Lighthouses*, 55.

5. Flint, *History of Lightships*, 4, 5.

6. Ibid., 8–9; Holland, *America's Lighthouses*, 56.

7. Flint, *Lightships of the U.S.*

8. Ibid.
9. Holland, *America's Lighthouses,* 57–58.
10. Flint, *Lightships of the U.S.*
11. Holland, *America's Lighthouses,* 58.
12. Flint, *Lightships of the U.S.*
13. Ibid.
14. Nalty and Strobridge, "Bright and Steadfast Light," 39; Flint, *Lightships of the U.S.*
15. Truman R. Strobridge, "Early Coast Guard Lightships on the Great Lakes," *Inland Seas: Quarterly Journal of the Great Lakes Historical Society* 29, no. 1 (Spring 1973): 16.
16. U.S. Lighthouse Board, *Annual Report of U.S. Light-House Board to the Secretary of the Treasury for the Fiscal Year Ending June 30, 1891* (Washington, D.C.: U.S. Government Printing Office, 1891), 27 (hereafter cited as *LHB, AR,* with appropriate year and page[s]).
17. Strobridge, "Great Lakes Lightships," 17.
18. Ibid., 17–18.
19. Flint, *Lightships of the U.S.*
20. Strobridge, "Great Lakes Lightships," 18.
21. Ibid., 17–18; *LHB AR, 1892,* 141.
22. Telegram, 12 November 1891, U.S. Lighthouse Board Letters to Ninth District Inspector, RG26, NAB.
23. Strobridge, "Great Lakes Lightships," 19–20.
24. *LHB AR, 1892,* 142.
25. 12 January 1892, Minutes of Meetings of the Light-House Board, RG26, NAB.
26. Quoted in Strobridge, "Great Lakes Lightships," 21.
27. Ibid., 23–25.
28. Holland, *America's Lighthouses,* 60.
29. Ibid., 61; Flint, *Lightships of the U.S.*
30. Holland, *America's Lighthouses,* 61.
31. *American State Papers, Class 4, Commerce and Navigation,* vol. 7 (Washington, D.C.: Gales and Seaton, 1832), (serial no. 14), 639, 690–91; *American State Papers, Class 4, Commerce and Navigation,* vol. 2 (Washington, D.C.: Gales and Seaton, 1834), (serial no. 15), 520–21.
32. Holland, *History of Cape Hatteras,* 31–32; Flint, *Lightships of the U.S.*
33. Holland, *History of Cape Hatteras,* 32–33.
34. Ibid., 33–34; *Dictionary of American Biography,* 649–50.
35. Holland, *History of Cape Hatteras,* 34–36; Flint, *Lightships of the U.S.*
36. Flint, *Lightships of the U.S.*
37. Ibid.
38. Putnam, *Lighthouses and Lightships,* 142–43.
39. Ibid.
40. Ibid.; quoted in James A. Gibbs, Jr., *Sentinels of the North Pacific: The Story of Pacific Coast Lighthouses and Lightships* (Portland, Ore.: Binfords and Mort, 1955), 173.
41. Flint, *Lightships of the U.S.* The fate of the ship after 1984 is not known.
42. Gustav Kobbe, "Life on the South Shoal Lightship" (Golden, Colo.: Outbooks, 1981, reprint), n.p.
43. Nalty and Strobridge, "Bright and Steadfast Light," 39.
44. Flint, *Lightships of the U.S.*; Holland, *America's Lighthouses,* 64.
45. Gibbs, *Sentinels of the North Pacific,* 173.
46. Quoted in ibid., 183.
47. Flint, *History of U.S. Lightships,* 17, 19; Flint, *Lightships of the U.S.*

48. Flint, *Lightships of the U.S.*

49. Ibid.

50. Quoted in Kobbe, "South Shoal Lightship," 35.

51. Quoted in Caldwell, *Lighthouses of Maine*, 234.

52. Holland, *America's Lighthouses*, 64; Nalty and Strobridge, "A Bright Steadfast Light," 39.

53. Gibbs, *Sentinels of the North Pacific*, 160–61; Nalty and Strobridge, "Bright Steadfast Light," 37.

54. Flint, *Lightships of the U.S.*

55. George R. Putnam, "New Safeguards for Ships in Fog and Storm," *National Geographic* 70, no. 2 (August 1936): 177.

56. Flint, *Lightships of the U.S.*

57. Nalty and Strobridge, "A Bright Steadfast Light," 39.

58. D. K. Robinson, "United States Coast Guard Offshore Light Stations," *U.S. Coast Guard Engineer's Digest* 140 (September–October 1961): 26–27.

59. Ibid., 27–28; Ronald D. Rosie, "The Offshore Light Station Program," *U.S. Coast Guard Engineer's Digest* (October–November–December 1967): 52. "We have demolished Brenton Reef tower [1992] and will demolish Buzzard Bay tower in the summer of 1996." Undated response to request from the author to the Historian of the U.S. Coast Guard and the Office of Aids to Navigation, U.S. Coast Guard Headquarters, Washington, D.C., by C. Mosher, Short Range Aids to Navigation, U.S. Coast Guard Headquarters.

60. J. W. Coste, Jr., and J. A. McIntosh, "Special Report: The Prototype LNB," *U.S. Coast Guard Engineer's Digest* 165 (October–November–December 1969): 35–36, 37.

61. Interview, C. Mosher, Short Range Aids to Navigation, U.S. Coast Guard Headquarters, Washington, D.C., 27 March 1996, with the author.

62. Quoted in Gibbs, *Sentinels of the North Pacific*, 175; quoted in Flint, *History of U.S. Lights*, 20; Flint, *Lightships of the U.S.*

7. THE BLACK FLEET

1. Putnam, *Sentinel of the Coasts* (New York: W. W. Norton, 1937), 258.

2. K. K. Cowart, "Development of Vessels Servicing Aids to Navigation for the U.S. Coast Guard," The Society of Naval Architects and Marine Engineers, *Transactions* 66 (1958): 517.

3. "Tenders and Light-Vessels," n.d., 98, a handwritten volume created by the U.S. Lighthouse Service on file, Historian.

4. Richard D. White and Truman R. Strobridge, "Nineteenth-Century Lighthouse Tenders on the Great Lakes," *Inland Seas: Quarterly Journal of the Great Lakes Historical Society* 31, no. 2 (Summer 1975): 91.

5. Light-House Board Journal, 14 November 1857, RG26, NAB.

6. "Tenders and Light-Vessels," 98.

7. Light-House Board Journal, 6 May 1862, RG26, NAB.

8. Ibid.

9. Ibid.

10. "Tenders and Light-Vessels," 98.

11. Ibid.

12. Ibid.

13. Unless otherwise noted, all material on the *Belle* is from Richard D. White and Truman R. Strobridge, "The *Belle:* Last Sailing Lighthouse Tender on the Great Lakes," *Telescope* 25, no. 2 (March–April 1976): 36–37.

14. *LHB AR, 1867,* 37.
15. "Tenders and Light-Vessels," 82.
16. *LHB AR, 1870,* 69.
17. *LHB AR, 1873,* 84.
18. *LHB AR, 1906,* 110; *LHB AR, 1876,* 50.
19. *LHB AR, 1894,* 167–68.
20. *LHB AR, 1906,* 110.
21. White and Strobridge, "Nineteenth-Century Lighthouse Tenders," 94–96.
22. U.S. Lighthouse Board, *Lighthouse Laws and Appropriations, 1789–1855* (Washington, D.C.: U.S. Government Printing Office, 1855), 169.
23. U.S. Lighthouse Board, *U.S. Lighthouse Board Report and Estimates 1856* (Washington, D.C.: U.S. Government Printing Office, 1856), 39.
24. Richard D. White, Jr., "Saga of the Side-Wheel Steamer *Shubrick:* Pioneer Lighthouse Tender of the Pacific Coast," *American Neptune* 36, no. 1 (1976): 46; "Tenders and Light-Vessels," 121.
25. "Tenders and Light-Vessels," 28.
26. White, "Saga of the *Shubrick,*" 47.
27. Ibid.
28. "Tenders and Light-Vessels," 121.
29. *LHB AR, 1857,* 12.
30. *Daily Alta California* (San Francisco), 28 May 1858, 2.
31. White, "The Saga of the *Shubrick,*" 48.
32. Ibid., 48–49.
33. *LHB AR, 1860,* 1.
34. Ibid., 8.
35. Smith did not live to see his plans fulfilled; he died in the sinking of the *Brother Jonathan* near Crescent City, California, on 30 July 1865. All material on Smith is found in Paul J. Martin and Peggy Brady, *Port Angeles, A History,* vol. 1 (Port Angeles, Wash.: Peninsula Publishing, 1983), 13–29.
36. White, "Saga of the *Shubrick,*" 50–51.
37. Quoted in ibid., 52.
38. Ibid.; "Tenders and Light-Vessels," 121.
39. White, "Saga of the *Shubrick,*" 52–53.
40. "Tenders and Light-Vessels," passim.
41. Cowart, "Development of Vessels," 517.
42. Ibid., 519.
43. Ibid., 519, 521. The tenders were *Crocus, Sunflower, Larkspur, Ivy,* and *Magnolia.*
44. Ibid., 521. The tenders were *Manzanita, Sequoia, Cypress, Orchid, Tulip, Hibiscus, Anenome,* and *Kukui.*
45. Ibid., 523–24. The tenders of the mine-laying class were *Lupine, Illex, Lotus, Speedwell,* and *Acacia.*
46. Ibid., 523–24.
47. White and Strobridge, "Nineteenth-Century Lighthouse Tenders," 94. The year of change from lighthouse tender to buoy tender was determined by examining the U.S. Coast Guard's *Standard Distribution List* publications, which give the names and addresses of all U.S. Coast Guard units by class; Historian.
48. Unless otherwise noted, all material on the *Columbine* rescue is found in Dennis L. Noble and Barbara Voulgaris, *Alaska and Hawaii: A Brief History of U.S. Coast Guard Operations* (Washington, D.C.: U.S. Coast Guard Historian's Office, 1991), 9.

49. Unless otherwise noted, all material on the rescue by Captain Smith is found in *USLSS AR, 1878,* 50.
50. Ralph Shanks, Jr., "Tenders: The Unsung Heroes," *Keeper's Log* (Winter 1987): 12.
51. There were six ships in the 180B, or *Mesquite,* class and twenty 180C, or *Iris,* class; all except the *Ironwood*—which was built at the U.S. Coast Guard Yard, Curtis Bay, Maryland—were completed at Duluth. Robert L. Scheina, *U.S. Coast Guard Cutters and Craft of World War II* (Annapolis, Md.: Naval Institute Press, 1982), 81, 92–100; Robert L. Scheina, *U.S. Coast Guard Cutters and Craft, 1946–1990* (Annapolis, Md.: Naval Institute Press, 1990), 141–55; Richard Fremont-Smith and David Pearl, "Present and Proposed Military Roles for Coast Guard Seagoing Buoy Tenders (WLBs)" (December 1982), 1–2, an unpublished position paper, Historian.
52. Richard D. White, "Buoy Tenders and Kamikazes," *U.S. Coast Guard Engineer's Digest* (July–August–September 1974): 27–28.
53. Ibid., 28; *The Coast Guard at War: Aids to Navigation,* vol. 15 (Washington, D.C.: Historical Section, Public Information Division, U.S. Coast Guard Headquarters, 1949), 82–83, 85.
54. Ibid., 29; ibid., 91–92. See also Richard D. White, Jr., "Aids to Navigation in Puerto Rico 1927–1942: History of the Buoy Tender Acacia," *Revista/Review Interamericana* 4, no. 4 (Winter 1974–75): 518–25.
55. White, "Buoy Tenders and Kamikazes," 29–30.
56. Ibid., 30.
57. Ibid., 30–31.
58. Amy K. Marshall, *A History of Buoys and Tenders* (Washington, D.C.: U.S. Coast Guard, 1996), 16. Marshall's short work is the first that has been written on the subject of buoys and buoy tenders in many years and is a good introduction to the subject; Historian.
59. White, "Buoy Tenders and Kamikazes," 27.

8. Fog Signals and Fancy Buoys

1. Holland, *America's Lighthouses,* 202; Putnam, *Sentinel of the Coast,* 199.
2. Strobridge, *Chronology of Aids to Navigation,* 4.
3. Johnson, *Modern Light-House Service,* 64–65; Wayne Wheeler, "The History of Fog Signals, Part 1," *Keeper's Log* (Summer 1990): 21.
4. Hans Christian Adamson, *Keepers of the Lights* (New York: Greenberg, 1955), 364.
5. Johnson, *Modern Light-house Service,* 65–66.
6. Clifford and Clifford, *Women Who Kept the Lights,* 50.
7. Ibid., 50–51; Wayne Wheeler, "History of Fog Signals, Part 1," *Keeper's Log* (Summer 1990): 23.
8. Adamson, *Keepers of the Lights,* 366–67.
9. Wayne Wheeler, "The History of Fog Signals, Part 2," *Keeper's Log* (Fall 1990): 8.
10. Quoted in ibid., 10.
11. Quoted in ibid., 11.
12. Holland, *America's Lighthouses,* 204.
13. Wayne Wheeler, "History of the Fog Signal, Part 2," *Keeper's Log* (Fall 1990): 13.
14. Ibid., 13.
15. Quoted in ibid.
16. Marshall, *A History of Buoys and Tenders,* 1; "Lecture on Buoys," unknown author and unknown date, but probably 1942–1945, 1, Historian.

17. Ibid., 1–2; Holland, *America's Lighthouses,* 206–7.

18. Ibid., 2.

19. Ibid., 2–3; Holland, *America's Lighthouses,* 207; Wayne Wheeler, "Buoys: Guideposts of the Sea," *Keeper's Log* (Fall 1986): 15.

20. Ibid., 3–4.

21. On 15 April 1982, "the United States agreed to make modifications [to the former buoy system] to incorporate the International Association of Lighthouse Authorities . . . Maritime Buoyage System." U.S. Coast Guard, *Light List,* vol. 1, *Atlantic Coast* (Washington, D.C.: U.S. Government Printing Office, 1984), xiii.

22. Ibid., xiv.

23. Ibid., xiii.

24. Holland, *America's Lighthouses,* 207–8.

25. Gibbs, *Sentinels of the North Pacific,* 95; Holland, *America's Lighthouses,* 208.

26. Holland, *America's Lighthouses,* 206; George R. Putnam, "New Safeguards for Ships in Fog and Storm," *National Geographic* 70, no. 2 (August 1936): 169.

27. Unless otherwise noted, all material on LORAN is found in H. R. Kaplan and James F. Hunt, *This Is the Coast Guard* (Cambridge, Md.: Cornell Maritime Press, 1972), 115–19; "Principles and History of LORAN System of Navigation," U.S. Coast Guard Academy *Alumni Association Bulletin* 7, no. 8 (November 1945): 274–80.

28. Weather does play some part in obtaining repeatability of positions. See Jesse J. Leaf, "LORAN Accuracy: Part 1," *Yachting* 166, no. 3 (September 1989): 38–39, 42.

29. Ibid., 38.

30. Ibid., 38–39, 42.

31. This is yet another subject on which a monograph should be written. Many of the LORAN stations were located in extremely isolated areas throughout the world. At one time, stories circulated throughout the U.S. Coast Guard regarding how different it was at these units, especially those stations in the Pacific Ocean area.

9. Eight Bells

1. *Aids to Navigation,* 10.

2. Photocopy of confidential letter, Col. George C. Van Dusen, Military Intelligence, to Col. W. Preston Corderman, as an enclosure of Navy Department, Division of Naval Intelligence, memorandum to commandant, U.S. Coast Guard (intelligence officer), dated 26 May 1942. Memorandum and attached letter are located in Lighthouse Miscellaneous file, Historian.

3. Malcolm F. Willoughby, *The U.S. Coast Guard in World War II* (Annapolis, Md.: U.S. Naval Institute Press, 1957), 126.

4. *Aids to Navigation,* 62–64.

5. Ibid., 22.

6. Ibid., 23.

7. Ibid., 54.

8. Ibid., 10; Willoughby, *Coast Guard in World War II,* 125.

9. *Aids to Navigation,* 3.

10. Gleason, *Kindly Lights,* 124.

Glossary of Nautical Terms

Aid to navigation Any device external to a vessel or aircraft specifically intended to assist navigators in determining their position or safe course, or to warn them of dangers or obstructions to navigation.

Airway beacon A rotating, reflecting light similar to that used at airports.

Astragals The vertical members of the metal frames in which the storm panes surrounding the lantern room sit.

Balcony The outside projecting walkway around a lantern room. Most lighthouses had one, and a few had two. Also known as a gallery.

Ball vent The round-shaped device that sits above the lantern room and through which air can pass. The ball vent was topped with a lightning rod and sometimes a wind vane.

Bark Usually a three-masted sailing vessel, square-rigged on the fore and main masts and fore-and-aft-rigged on the mizzen. *See* Masts.

Barkentine Three- or more masted, square-rigged on the fore mast and fore-and-aft-rigged on the main and mizzen, or any additional, masts. *See* Masts.

Bilge keel A fin fitted to the hull on each side of the ship at the turn of the bilge to reduce rolling.

Characteristic The specific audible, visual, or electronic signal displayed by an aid to navigation that assists in the identification of the aid. Characteristic refers to lights, sound signals, radio beacons, and day beacons. Light characteristics include *fixed,* in which the light shows continuously and steadily; *occulting,* in which the total duration of darkness and the intervals of darkness (eclipses) are usually of equal duration; and *flashing,* in which the total duration of light in a period is shorter than the total duration of darkness, and the appearance of light (flashes) are usually of equal duration.

Day mark The daytime identifier of an aid to navigation presenting one of several shapes (square, triangle, rectangle) and colors (red, green, white, orange, yellow, or black).

Faked A method of laying out line (rope) or chain in a series of lengths, instead of coiling, so that it will run free when pulled.

Fixed light *See* Characteristic.

Flashing light *See* Characteristic.

Focal plane A plane that is level with the plane of light that passes through the principal focus of the lens and is measured by the distance above mean level.

Gallery *See* Balcony.

Hook Nautical slang for anchor.

Lamp room The glass-enclosed room atop a lighthouse that contains the lens. Also called lantern room and lanthorn room. For many years this room was open to the elements.

Lantern room *See* Lamp room.

Lanthorn room *See* Lamp room.

Light A lamp or lantern. Sometimes used to mean a light station or lighthouse. *See also* Lighthouse, Light station.

Lighthouse A tower or building with a light for guiding ships. *See also* Light, Light station.

Light station One or more buildings on a manned light site.

Line Nautical term for rope.

Masts On three-masted ships, the masts are—after the bow—fore, main, and mizzen.

Mushroom anchor A rounded anchor used on lightships for its holding abilities.

Occulting light *See* Characteristic.

Plank owner Nautical term for a crew member who is aboard a ship when it is commissioned.

Radio beacon A radio-sending device transmitting a code by which a navigator can determine the position of a ship.

Storm panes The glass windows that enclose the lantern room.

Swallow the hook Term used to describe a sailor leaving the sea to live ashore.

Turtle back deck A weather deck on the forecastle or poop which is rounded over at the sides to shed water in heavy sea.

Watchroom At a light station, the room where the keeper stands watch, usually just below the lantern room or beside the tower entrance.

SELECTED BIBLIOGRAPHY

ARCHIVAL SOURCES

Records of the U.S. Coast Guard. Record Group 26. National Archives, Washington, D.C.
"Descriptive List of Lighthouse Stations"
"Journal of the Light-House Board" (various dates)
"Journals," or logbooks (various dates)
"Lighthouse Annual Reports" (various dates)
"Lighthouse Clipping Files" (various dates)
"Lighthouse Correspondence" (various dates)
"Lighthouse Deeds and Contracts"
"Lighthouse Letters" (various dates)
"Lighthouse Site Files"
"Medals, U.S. Life-Saving Service"

U.S. Coast Guard Academy Museum, New London, Conn.
"The Lighthouse Service, 1845." Collection MVF 178.

U.S. Coast Guard Headquarters, Washington, D.C.
"Historian of the U.S. Coast Guard Lighthouse Files"

U.S. Lighthouse Society, San Francisco, Calif.
"Clifford Gallant Papers"
"Light Station Files"

Government Publications and Reports. Washington, D.C.

American State Papers, Class 4, vol. 1, *Commerce and Navigation, 1st–13th Congress.* Washington, D.C.: U.S. Government Printing Office, 1832.
U.S. Coast Guard. *The Coast Guard at War: Aids to Navigation,* vol. 15. Washington, D.C.: Historical Section, Public Information Division, U.S. Coast Guard Headquarters, 1949.
_____. *Historically Famous Lighthouses.* Washington, D.C.: U.S. Government Printing Office, 1986.

———. *A History of Blacks in the U.S. Coast Guard from 1790.* Washington, D.C.: U.S. Coast Guard, n.d.

———. *Some Unusual Incidents in Coast Guard History.* Washington, D.C.: Historical Section, Public Information Division, U.S. Coast Guard Headquarters, 1950.

U.S. Congress. Letter from the Secretary of the Treasury, Transmitting a Report of the Navy Commissioners of Their Proceedings Under the Act of 3d March 1837, Making Appropriations for Light-houses, etc., etc., 22 December 1837. H. Doc. 41, 25th Cong., 2d sess., 1837.

U.S. Congress. House. Letter from the Secretary of the Treasury, Transmitting the Information Required by a Resolution of the House of Representatives in Relation to the Best Mode of Managing the Light-House Establishment. H. Doc. 66, 24th Cong., 1st sess., 1838.

U.S. Congress. Senate. Report from the Secretary of the Treasury in Compliance with a Resolution of the Senate of the 25th Inst., Transmitting Copies of the Representations Made to Him Relative to the Light-houses of the United States, by the Messrs. Blunt, of New York, etc., 26 January 1838. S. Doc. 138, 25 Cong., 2 sess., 1839.

U.S. Congress. Report of the Light-House Establishment, J. C. Clark, Commissioner of Commerce. 28 May 1842. H. Rpt. 811, 24th Cong., 2 sess., 1842.

U.S. Congress. Letter from the Secretary of the Treasury, Transmitting a Report of the Fifth Auditor in Relation to the Light-houses, etc., 10 January 1844. H. Doc. 38, 28th Cong., 1st sess., 1844.

U.S. Congress. House. Letter from the Secretary of the Treasury, Transmitting the Fifth Auditor's Report Relative to Light-houses and Lights, 19 January 1844. H. Doc. 62, 28th Cong., 1st sess., 1844.

U.S. Congress. Report of the Offices Constituting the Light-House Board, Convened Under the Instructions from the Secretary of the Treasury to Inquire into the Condition of the Light-House Establishment of the United States, Under the Act of 3 March 1851. H. Ex. Doc. 55, 32d Cong., 1st sess., 1851.

U.S. Department of Commerce. *Lighthouse Bulletin.* Various dates. Washington, D.C.: U.S. Government Printing Office. Various dates.

U.S. Life-Saving Service. *Annual Report of the Operations of the United States Life-Saving Service.* Washington, D.C.: U.S. Government Printing Office, various dates.

U.S. Light-House Board. *Annual Report of the Light-House Board to the Secretary of the Treasury.* Washington, D.C.: U.S. Government Printing Office. Various dates.

———. *Compilation of Public Documents and Extracts from Reports and Papers Relating to Light-houses, Light-vessels, and Illuminating Apparatus, and to Beacons, Buoys, and Fog Signals, 1789 to 1871.* Washington, D.C.: U.S. Government Printing Office, 1871.

———. *Light-House Board Report and Estimates, 1856.* Washington, D.C.: U.S. Government Printing Office, 1856.

———. *Light-House Laws and Appropriations, 1789–1855.* Washington, D.C.: U.S. Government Printing Office, 1855.

U.S. Light-House Establishment. *Instructions to Keepers.* Washington, D.C.: U.S. Government Printing Office, various dates.

Newspapers

The Alaskan (Sitka)
Daily Alta California (San Francisco)
The Journal (Boston)
New York *Times*

BOOKS

Adamson, Hans Christian. *Keepers of the Lights*. New York: Greenberg, 1955.

Albee, Peggy A., Regina Binder, Millan Garland, and Larry Lowenthal. *Highland Lighthouse and Keeper's Dwelling: Historic Structure Report: North Truro, Massachusetts*. Lowell, Mass.: National Park Service, North Atlantic Region, 1994.

Baker, William A. *A History of the Boston Marine Society, 1742–1967*. Boston: Boston Marine Society, 1968.

Boutry, Georges A. *Augustin Fresnel: His Time, Life, and Work, 1788–1827*. London: John Murray, 1949.

Caldwell, Bill. *Lighthouses of Maine*. Portland, Me.: Gannett Books, 1986.

Carr, Dawson. *The Cape Hatteras Lighthouse: Sentinel of the Shoals*. Chapel Hill: University of North Carolina Press, 1991.

Carse, Robert. *Keepers of the Lights: A History of American Lighthouses*. New York: Charles Scribner's Sons, 1969.

Cipra, David L. *Lighthouses and Lightships of the Northern Gulf of Mexico*. Washington, D.C.: U.S. Coast Guard, 1983.

Clifford, Candace. *Inventory of Historic Light Stations*. Washington, D.C.: National Park Service, 1994.

Clifford, Mary Louise, and J. Candace Clifford. *Women Who Kept the Lights: An Illustrated History of Female Lighthouse Keepers*. Williamsburg, Va.: Cypress Communications, 1993.

Crowninshield, Mary Bradford. *All Among the Lighthouses on the Cruise of the Goldenrod*. Boston: Lothrop Publishing, 1886.

Dean, Love. *The Lighthouses of Hawai'i*. Honolulu: University of Hawaii Press, 1991.

DeWire, Elinor. *Guardians of the Lights: The Men and Women of the U.S. Lighthouse Service*. Sarasota, Fla.: Pineapple Press, 1995.

_____. *The Lighthouse Keeper's Scrapbook*. Gales Ferry, Conn.: Sentinel Publications, 1995.

Engel, Norma. *Three Beams of Light: Chronicles of a Lightkeeper's Family*. San Diego, Calif.: Tecolate Publications, 1986.

Evans, Stephen H. *The United States Coast Guard, 1790–1915: A Definitive History*. Annapolis, Md.: Naval Institute Press, 1949.

Flint, Willard. *A History of U.S. Lightships*. Washington, D.C.: U.S. Coast Guard, 1993.

_____. *Lightships of the U.S. Government: Reference Notes*. Washington, D.C.: U.S. Coast Guard, 1989.

Gibbs, James A., Jr. *Sentinels of the North Pacific: The Story of Pacific Coast Lighthouses and Lightships*. Portland, Ore.: Binfords and Mort, 1955.

_____. *Tillamook Light*. Portland, Ore.: Binford and Mort, 1953.

_____. *West Coast Lighthouses: A Pictorial History of the Guiding Lights of the Sea*. Seattle, Wash.: Superior Publishing, 1974.

Gleason, Sarah C. *Kindly Lights: A History of Lighthouses of Southern New England*. Boston: Beacon Press, 1991.

Hall, Pat. *The Point Reyes Light*. Point Reyes, Calif.: Coastal Parks Association, 1979.

Hamilton, Harlan. *Lights and Legends: A Historical Guide to Lighthouses of Long Island Sound, Fishers Island Sound and Block Island Sound*. Stamford, Conn.: Westcott Cove Publishing, 1987.

Holland, F. Ross, Jr. *America's Lighthouses: An Illustrated History*. Mineola, N.Y.: Dover Publications, 1988. (Originally published by Stephen Greene Press of Brattleboro, Vt., in 1972.)

_____. *The Aransas Pass Light Station: A History*. Corpus Christi, Tex.: privately printed, 1976.

_____. *A History of the Cape Hatteras Light Station: Cape Hatteras National Seashore, North Carolina.* Washington, D.C.: National Park Service, Division of History, 1968.

Howe, M. A. DeWolfe. *The Humane Society of the Commonwealth of Massachusetts: An Historical Review.* Boston: Boston Humane Society, 1918.

Johnson, Arnold Burges. *The Modern Light-House Service.* Washington, D.C.: U.S. Government Printing Office, 1890.

Johnson, Robert Erwin. *Guardians of the Sea: History of the United States Coast Guard, 1915 to the Present.* Annapolis, Md.: Naval Institute Press, 1987.

Jones, Dorothy Holder, and Ruth Sexton Sargent. *Abbie Burgess, Lighthouse Heroine.* New York: Funk and Wagnalls, 1969.

Kaplan, H. R., and James F. Hunt. *This is the Coast Guard.* Cambridge, Md.: Cornell Maritime Press, 1972.

Kobbe, Gustav. *Life on the South Shoal Lightship.* (Reprinted by Outbooks of Golden, Colo., in 1981.)

Kozma, LuAnne Gaykowski. *Living at a Lighthouse: Oral Histories from the Great Lakes.* Detroit, Mich.: Great Lakes Lighthouse Keepers Association/Harlo Printing, 1987.

Livingston, Dewey, and David Snow. *The History and Architecture of Point Reyes Light Station.* Point Reyes, Calif.: National Park Service, 1990.

Lowery, Shannon. *Northern Lights: Tales of Alaska's Lighthouses and Their Keepers.* Harrisburg, Pa.: Stackpole Books, 1992.

Marshall, Amy K. *A History of Buoys and Tenders.* Washington, D.C.: U.S. Coast Guard, 1996.

Martin, Paul J., and Peggy Brady. *Port Angeles, A History,* vol. 1. Port Angeles, Wash.: Peninsula Publishing, 1983.

Melville, David. *An Expose of Facts, Respectfully Submitted to the Government and Citizens of the United States, Relating to the Conduct of Winslow Lewis, of Boston, Addressed to the Hon. the Secretary of the Treasury.* Providence, R.I.: Miller and Hutchens, 1819.

Morgan, William James, et al. *Autobiography of Rear Admiral Charles Wilkes, U.S. Navy, 1798–1877.* Washington, D.C.: U.S. Department of the Navy, 1978.

Naish, John. *Seamarks: Their History and Development.* London: Stanford Maritime, 1985.

Noble, Dennis L. *The Coast Guard in the Pacific Northwest.* Washington, D.C.: U.S. Coast Guard, 1988.

_____. *Southwest Pacific: A Brief History of U.S. Coast Guard Operations.* Washington, D.C.: U.S. Coast Guard, 1989.

_____. *That Others Might Live: The U.S. Life-Saving Service, 1878–1915.* Annapolis, Md.: Naval Institute Press, 1994.

_____, and T. Michael O'Brien. *Sentinels of the Rocks: From "Graveyard Coast" to National Lakeshore.* Marquette: Northern Michigan University Press, 1979.

_____, and Barbara Voulgaris. *Alaska and Hawaii: A Brief History of U.S. Coast Guard Operations.* Washington, D.C.: U.S. Coast Guard, 1991.

Putnam, George R. *Lighthouses and Lightships of the United States.* Boston: Houghton and Mifflin Company, 1917.

_____. *Sentinel of the Coasts: The Log of a Lighthouse Engineer.* New York: W. W. Norton, 1937.

Roberts, Bruce, and Ray Jones. *Western Lighthouses: Olympic Peninsula to San Diego.* Oldsaybrook, Conn.: Globe Pequot Press, 1993.

Scheina, Robert L. "The Evolution of the Lighthouse Tower," in *Lighthouses: Then and Now.* Washington, D.C.: U.S. Coast Guard, n.d.

_____. *U.S. Coast Guard Cutters and Craft of World War II.* Annapolis, Md.: Naval Institute Press, 1982.

_____. *U.S. Coast Guard Cutters and Craft, 1946–1990.* Annapolis, Md.: Naval Institute Press, 1990.

Shanks, Ralph C., Jr., and Janetta Thompson Shanks. *Lighthouses and Lifeboats on the Redwood Coast*. San Anselmo, Calif.: Costano Books, 1978.

Snow, Edward Rowe. *The Lighthouses of New England*. New York: Dodd, Mead, 1973.

_____. *Famous New England Lighthouses*. Boston: Yankee Publishing, 1945.

Stevenson, D. Alan. *The World's Lighthouses Before 1820*. London: Oxford University Press, 1959.

Stick, David. *North Carolina Lighthouses*. Raleigh: North Carolina Department of Cultural Resources, Division of Archives and History, 1980.

Strobridge, Truman R. *Chronology of Aids to Navigation and the Old Lighthouse Service, 1716–1939*. Washington, D.C.: U.S. Coast Guard, 1974.

Weiss, George. *The Lighthouse Service: Its History, Activities and Organization*. Baltimore: Johns Hopkins Press, 1926.

Willoughby, Malcolm F. *The U.S. Coast Guard in World War II*. Annapolis, Md.: Naval Institute Press, 1957.

Whitney, Dudley. *The Lighthouse*. Boston: New York Graphic Society, 1975.

ARTICLES

Anonymous. "Principles and History of LORAN System of Navigation." *U.S. Coast Guard Academy Alumni Association Bulletin* 7, no. 8 (November 1945): 274–80.

Baker, William A. "U.S. Light-Vessel *No. 50* Columbia River." *American Neptune*, Vol. 9, No. 4 (October 1949): 273–77.

Bergin, Layne. "Screwpile Lighthouses: From Britain to the Bay." *Keeper's Log* (Spring 1987): 10–15.

Bretherton, Bernard J. "The Destruction of Birds by Lighthouses," *Osprey* (1902): 76–78.

Cairo, Robert F. "Notes on Early Lightship Development." *U.S. Coast Guard Engineer's Digest* 188 (July–August–September 1975): 3–14.

Churchill, Sam. "The Day 'Terrible Tilly's' Light Nearly Died in a Sea of Terror." *Northwest Magazine* (3 December 1972): 6, 8.

Coste, J. W., Jr., and J. A. McIntosh. "Special Report: The Prototype LNB." *U.S. Coast Guard Engineer's Digest* 165 (October–November–December 1969): 35–36, 37.

Cowart, K. K. "Development of Vessels Servicing Aids to Navigation for the U.S. Coast Guard." *Transactions* 66 (1958): 516–78.

DeWire, Elinor. "The Bulging Eye of Makapu'u." *Keeper's Log* (Summer 1986): 2–5.

_____. "Specters on the Spiral Stairs." *Keeper's Log* (Winter 1986): 8–11.

_____. "Women of the Lights." *American History Illustrated* 21, no. 10 (February 1987): 42–48.

Fraser, Robert. "I.W.P. Lewis: Father of America's Lighthouse System." *Keeper's Log* (Winter 1989): 8–10.

Gallant, Clifford. "Emily Fish: The Socialite Keeper." *Keeper's Log* (Spring 1985): 8–13.

Gibbs, James. "Terrible Tilly Revisited." *Keeper's Log* (Summer 1988): 12–17.

Harris, Patricia. "Michigan City, Indiana's Only Lighthouse." *Keeper's Log* (Spring 1987): 22–25.

Kern, Florence. "Lighthousing in the 1890s." *U.S. Coast Guard Academy Alumni Association Bulletin* 40, no. 6 (November–December 1978): 33–38.

Leaf, Jesse J. "LORAN Accuracy: Part 1." *Yachting* 166, no. 3 (September 1989): 38, 40, 42.

_____. "LORAN Accuracy: Part 2." *Yachting* 166, no. 4 (October 1989): 55–57.

Lindquist, Lindy. "L'Homme De Genie: The Unknown Tale of How 'Fresnel's Fancies' Lit the Lighthouses of the World." *Oceans* (November 1974): 58–63.

Livingston, Dewey. "The Keepers of the Light." *Keeper's Log* (Winter 1987): 16–19.

_____. "Point Reyes." *Keeper's Log* (Winter 1991): 2–15.

Macy, Robert H. "The Consolidation of the Lighthouse Service with the Coast Guard," U.S. Naval Institute *Proceedings* 66, no. 1 (January 1940): 58–71.

McClelland, C. H. "Terrible Tilly." *Keeper's Log* (Summer 1987): 2–9.

Nalty, Bernard C., and Truman R. Strobridge. "A Bright and Steadfast Light." *U.S. Coast Guard Academy Alumni Bulletin* 38, no. 6 (November–December 1975): 37–41.

Noble, Dennis L., and Ralph E. Eshelman. "A Lighthouse for Drum Point." *Keeper's Log* (Spring 1987): 2–9.

Pedersen, Ted, as told to Ed Moreth. "Ted Pedersen: An Alaskan Lighthouse Keeper." *Keeper's Log* (Spring 1990): 16–19.

Putnam, George R. "Beacons of the Sea: Lighting the Coasts of the United States." *National Geographic* 24, no. 1 (January 1913): 1–53.

———. "New Safeguards for Ships in Fog and Storm." *National Geographic* 70, no. 2 (August 1936): 169–200.

Ridgely-Nevitt, Cedric. "A Light-vessel of 1823 built by Henry Eckfor." *American Neptune* 5, no. 2 (April 1945): 115–20.

Robinson, D. K. "United States Coast Guard Offshore Light Stations." *U.S. Coast Guard Engineer's Digest* 140 (September–October 1961): 26–29.

Rosie, Ronald D. "The Offshore Light Station Program." *U.S. Coast Guard Engineer's Digest* (October–November–December 1967): 50–55.

Rozin, Skip. "Who Mourns the Vanishing Wickies?" *Audubon* (May 1972): 30–35.

Shanks, Ralph, Jr. "Tenders: The Unsung Heroes." *Keeper's Log* (Winter 1987): 12–15.

Strobridge, Truman R. "Blacks and Lights." *Shipmates* 3, no. 4 (Summer 1975): 14–17, 31.

———. "Early Coast Guard Lightships on the Great Lakes." *Inland Seas: Quarterly Journal of the Great Lakes Historical Society* 29, no. 1 (Spring 1973): 16–25.

Thompson, Sue Ellen. "The Light Is My Child," *Log of Mystic Seaport* 32, no. 3 (Fall 1980): 90–98.

Updike, Richard W. "Winslow Lewis and the Lighthouses." *American Neptune* 28, no. 1 (January 1968): 31–48.

———. "Augustin Fresnel and his Lighthouse Lenses." *Log of Mystic Seaport* 19, no. 2 (Spring/Summer 1967): 34–39.

Wheeler, Wayne. "Augustin Fresnel and His Magic Lantern." *Keeper's Log* (Winter 1985): 8–10.

———. "Buoys: Guideposts of the Sea." *Keeper's Log* (Fall 1986): 10–15.

———. "The History of Fog Signals, Part 1." *Keeper's Log* (Summer 1990): 20–23.

———. "The History of Fog Signals, Part 2." *Keeper's Log* (Fall 1990): 8–15.

———. "The Keeper's New Clothes." *Keeper's Log* (Summer 1985): 10–13.

White, Richard D. "Aids to Navigation in Puerto Rico 1927–1942: History of the Buoy Tender Acacia." *Revista/Review Interamericana* 4, no. 4 (Winter 1974–75): 518–25.

———. "Buoy Tenders and Kamikazes." *U.S. Coast Guard Engineer's Digest* (July–August–September 1974): 27–31.

———. "Destination Nowhere: The Twilight of the Lightship." U.S. Naval Institute *Proceedings* 102 (March 1976): 64–69.

———. "Saga of the Side-Wheel Steamer *Shubrick:* Pioneer Lighthouse Tender of the Pacific Coast." *American Neptune* 36, no. 1 (1976): 45–53.

———, and Truman R. Strobridge. "The Belle: Last Sailing Lighthouse Tender on the Great Lakes." *Telescope* 25, no. 2 (March–April 1976): 36–37.

———, and Truman R. Strobridge. "Nineteenth-Century Lighthouse Tenders on the Great Lakes." *Inland Seas: Quarterly Journal of the Great Lakes Historical Society* 31, no. 2 (Summer 1975): 87–96.

INDEX

ABOUT THE AUTHOR

Dennis L. Noble served in the U.S. Coast Guard at a number of shore stations and in a variety of cutters, including the *Westwind, Northwind,* and *Edisto.* He made two trips to the Antarctic and six to the Arctic. He retired from the U.S. Coast Guard as a senior chief marine science technician in 1978.

Mr. Noble holds a Ph.D. in U.S. history from Purdue University. Since retirement from the Coast Guard he has worked as a park ranger, as a historian for the U.S. Army, as a public librarian, and as a librarian in a closed security prison. He is the author or co-author of five books, including *Wrecks, Rescues, and Investigations; Sentinels of the Rocks; The Eagle and the Dragon: The U.S. Military in China, 1901–1937; Forgotten Warriors: Combat Art from Vietnam,* and *That Others Might Live: The U.S. Life-Saving Service, 1878–1915* (the latter published by the Naval Institute Press in 1994). Dr. Noble has also written more than twenty articles for such periodicals as *Keeper's Log* and *Naval History.* He now resides in Sequim, Washington, when not on research trips.